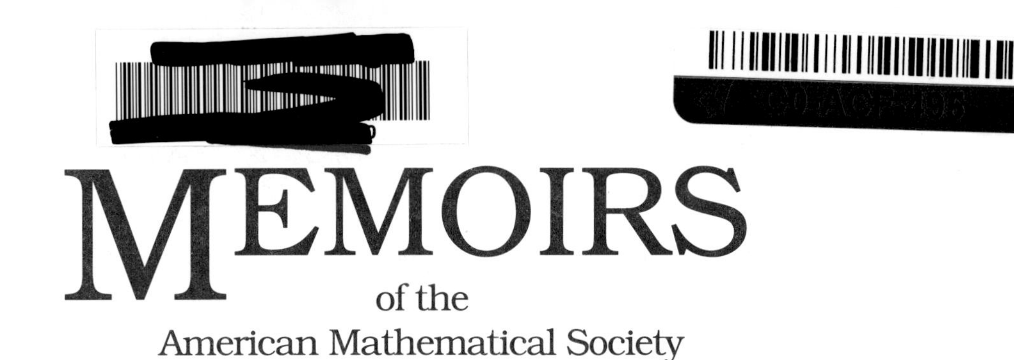

MEMOIRS
of the
American Mathematical Society

Number 455

Differential Operators and Highest Weight Representations

Mark G. Davidson
Thomas J. Enright
Ronald J. Stanke

November 1991 • Volume 94 • Number 455 (first of 4 numbers) • ISSN 0065-9266

American Mathematical Society
Providence, Rhode Island

1980 *Mathematics Subject Classification* (1985 *Revision*).
Primary 22E45, 22E47; Secondary 20G05, 32M15.

Library of Congress Cataloging-in-Publication Data

Davidson, Mark G., 1955–
Differential operators and highest weight representations/Mark G. Davidson, Thomas J. Enright, Ronald J. Stanke.
p. cm. – (Memoirs of the American Mathematical Society, ISSN 0065-9266; no. 455)
Includes bibliographical references and index.
ISBN 0-8218-2509-7
1. Semisimple Lie groups. 2. Representations of groups. 3. Differential operators. I. Enright, Thomas J. II. Stanke, Ronald J., 1954–. III. Title. IV. Series.
QA3.A57 no. 455
[QA387]
510 s–dc20 91-27804
[512′.55] CIP

Subscriptions and orders for publications of the American Mathematical Society should be addressed to American Mathematical Society, Box 1571, Annex Station, Providence, RI 02901-1571. *All orders must be accompanied by payment.* Other correspondence should be addressed to Box 6248, Providence, RI 02940-6248.

SUBSCRIPTION INFORMATION. The 1991 subscription begins with Number 438 and consists of six mailings, each containing one or more numbers. Subscription prices for 1991 are $270 list, $216 institutional member. A late charge of 10% of the subscription price will be imposed on orders received from nonmembers after January 1 of the subscription year. Subscribers outside the United States and India must pay a postage surcharge of $25; subscribers in India must pay a postage surcharge of $43. Expedited delivery to destinations in North America $30; elsewhere $82. Each number may be ordered separately; *please specify number* when ordering an individual number. For prices and titles of recently released numbers, see the New Publications sections of the NOTICES of the American Mathematical Society.

BACK NUMBER INFORMATION. For back issues see the AMS Catalogue of Publications.

MEMOIRS of the American Mathematical Society (ISSN 0065-9266) is published bimonthly (each volume consisting usually of more than one number) by the American Mathematical Society at 201 Charles Street, Providence, Rhode Island 02904-2213. Second Class postage paid at Providence, Rhode Island 02940-6248. Postmaster: Send address changes to Memoirs of the American Mathematical Society, American Mathematical Society, Box 6248, Providence, RI 02940-6248.

Printed in the United States of America.
The paper used in this book is acid-free and falls within the guidelines established to ensure permanence and durability. ♾

10 9 8 7 6 5 4 3 2 1 95 94 93 92 91

Table of Contents

Abstract

This memoir concerns the representation theory of semisimple Lie groups. The main results offer characterizations of unitary highest weight representations as solutions to systems of differential operators.

Key words and phrases: semisimple Lie groups and Lie algebras, Hermitian symmetric space, unitary representations, highest weight representations, Howe dual pairs, oscillator representation.

Received by the editor March 15, 1990 and in revised form December 17, 1990.

§0. Introduction and Summary of Results

From the algebraic perspective the theory of unitarizable highest weight modules is highly developed. The classification was given in 1981 ([15] and [27]) and more recently even the character and nilpotent cohomology formulas have been determined for G of classical type [12]. However, from the analytic point of view as originally presented by Harish-Chandra (cf. [18]) unitarizable highest weight modules occur as subspaces of certain spaces of vector valued polynomials; or equivalently, as subspaces of spaces of holomorphic sections for vector bundles on G/K. The basic themes of this memoir all concern the characterizations of these subspaces.

Let (G,K) be a Hermitian symmetric pair. We begin by associating to each unitary highest weight representation (π, L) of G a finite system $\mathcal{D}_L$ of polynomial differential operators. The correspondence allows the recovery of the infinitesimal representation as the kernel of the system. In this setting the full space of differential operators carries a natural action of K and the system $\mathcal{D}_L$ is minimal in that $\mathcal{D}_L$ spans an irreducible representation of K. Further, we can identify this representation

---

First author supported in part through a Fulbright Fellowship and the Science Institute of the University of Iceland, Reykjavik. Supported also in part by the Louisiana Education Quality Support Fund 86-LBR-016-04.

Second author supported in part by NSF grant #DMS-8902425.

Third author supported in part by a research grant from Baylor University.

In addition all three authors have been supported through NSF workshops organized by Hugo Rossi and conducted at University of Utah, Salt Lake City.

of K as the first cohomology $H^1(\mathfrak{p}_+,L)$ where $\mathfrak{p}_+$ is a Lie subalgebra of the Lie algebra of G, defined below, and give two separate rather explicit formulas for the highest weight of this representation of K. Our first characterization of $\mathcal{D}_L$ is given in Theorems 2.9 and 3.1 of the article. Later in this introduction we shall return to these issues in more detail and comment on various connections with the work of other authors, most notably H. P. Jakobsen and B. Kostant.

The second set of results contained here concerns what we call the Cone Decomposition. Let Λ_r denote the set of highest weights of unitarizable highest weight modules which are not isomorphic to modules induced from the proper parabolic subalgebra containing the Lie algebra, $\mathfrak{k}$, of K. We call Λ_r the set of reduction points. Clearly these are the modules of interest from our point of view since they are precisely those for which the system $\mathcal{D}_L$ is not empty. In section six we observe that Λ_r is the disjoint union of a finite set of cones, say $\{\Lambda_a\}$, $a\in\mathcal{A}$. Letting λ_a denote the vertex of the cone Λ_a we direct our attention to ways in which the vertex λ_a determines the structure of all the modules parameterized by the points in the cone Λ_a. If L has highest weight λ we write $\mathcal{D}_\lambda$ in place of $\mathcal{D}_L$. Then the first result along these lines shows that, for any $\lambda\in\Lambda_a$, the system $\mathcal{D}_\lambda$ is obtained as a "shift" of $\mathcal{D}_{\lambda_a}$. Theorem 6.15 makes this notion of shift precise and is called the Factorization Theorem.

Connections between the cone decomposition and the representation theory are quite surprising and more explicit in the setting of harmonic polynomials and the oscillator representation. The setup here is described in section seven. We refer to it as the harmonic setting. Here the shift mentioned above is given explicitly by a multiplication of polynomials.

There are two other main results appearing in the harmonic setting. The first is related to the cone decomposition. For a cone Λ_a we define a finite system $\mathcal{F}_a$ of differential operators which determines through its kernel all of the relevent modules with highest weights in the cone Λ_a.

In the second we associate to a unitarizable highest weight module L an affine variety V and then recover L as the image in the coordinate ring of V of a certain space of harmonic-valued polynomials. We find that V is the graph of a map θ and the image of θ is the closure of a single $K_{\mathbb{C}}$-orbit in the complexified Lie algebra of G.

Following this we offer explicit bases for the systems $\mathcal{F}_a$. The answers here are expressed in terms of both determinants and Pfaffians of certain block matrices. Finally as illustrative examples of what has been developed we give characterizations of both the ladder and Wallach representations.

We now turn to a somewhat more detailed account of the main results. The reader may wish to first view the table of contents as an outline of what is to come. Let $\mathfrak{g}$ (respectively $\mathfrak{k}$) be the complexified Lie algebra of G (resp. K) and let $\mathfrak{g} = \mathfrak{k} \oplus \mathfrak{p}$ be a Cartan decomposition of $\mathfrak{g}$. Then $\mathfrak{p}$ decomposes into two irreducible K-submodules $\mathfrak{p} = \mathfrak{p}_+ \oplus \mathfrak{p}_-$, each is an abelian subalgebra of $\mathfrak{g}$. The characterization of $\mathcal{D}_L$ involves both an analytic as well as an algebraic part. The analytic component is the theory of reproducing kernels which gives an explicit nondegenerate pairing of a generalized Verma module N and its conjugate dual module N*. The pairing appears explicitly through differentation. The upshot is Theorem 2.9 which asserts that any set of generators (over the enveloping algebra of $\mathfrak{p}_-$) for the maximal submodule J of N will give the system $\mathcal{D}_L$. Then Theorem 3.1 sharpens this result by giving a canonical choice for $\mathcal{D}_L$, proving it is an irreducible representation of K and is equivalent to $H^1(\mathfrak{p}_+, L)$. The irreducibility is established by using the cohomology formulas established in [12] for G of classical type. Since we need these formulas in general, section four of this article is devoted to the formulation and proofs of these formulas for the exceptional cases (cf. Definition 4.2 and Theorem 4.1). In doing this we generalize Collingwood's notion of a Kostant module [8] to the setting of singular infinitesimal character and then prove that every unitarizable highest weight module is a Kostant module. The first formula for the highest weight of $H^1(\mathfrak{p}_+, L)$ occurs in this context and is given as (3.8).

The second part of the article describes the Cone Decomposition (6.6) and the Factorization Theorem (6.15) in the general setting and then turns to various specializations, usually in the harmonic setting. The proof of the Factorization Theorem uses a second formula for the highest weight of $H^1(\mathfrak{p}_+, L)$. This formula is established by the second author and A. Joseph [13] and presents this representation of K as the Parthasarathy-Ranga Rao-Varadarajan (PRV) component in a special tensor product. This formula is given as Proposition 6.8 here. Our earlier proof of the Factorization Theorem was independent of this formula but somewhat longer.

In section seven we introduce the oscillator representation, harmonic polynomials and the Howe theory of dual reductive pairs. Let H_λ be an irreducible $\mathfrak{k}$-module with highest weight λ whose underlying space is a space of harmonic polynomials on the vector space M. Then the generalized Verma module $N(\lambda+\rho) = U(\mathfrak{g}) \otimes_{U(\mathfrak{k}\oplus\mathfrak{p}_+)} H_\lambda$ can be viewed either as a space of H_λ-valued polynomials on $\mathfrak{p}_+$ or as a space of scalar valued polynomials on $\mathfrak{p}_+ \times M$. Of course it is this second set of variables which plays such a strong role in the harmonic setting. The variety V mentioned above is the graph of a map θ from M into $\mathfrak{p}_+$. This result is given as (7.7).

Let Ω denote the subset of Λ_r of those points whose corresponding representations are realized in the harmonic setting. It was conjectured by Kashiwara and Vergne [31] and later proven in [14] that Ω equals Λ_r for SU(p,q) or the double cover of $Sp(n,\mathbb{R})$. In the classification article [15] it was noted in passing that Ω equals Λ_r for SO*(2m) as well. This is not the case and part of section seven addresses the precise relationship between these two sets.

Section eight begins with a generalization and application of the theory of Young products to the setting of harmonic polynomials. These results are then applied to give the refinement of the Factorization Theorem mentioned above. Here the main result is (8.4) and its corollary.

In section nine the role of the variables of M enter in the most decisive way. The systems $\mathcal{D}_\lambda$ are systems of differential equations in the variables of $\mathfrak{p}_+$. However, for a cone Λ_a in the harmonic setting we define a finite system $\mathcal{F}_a$ of differential operators in the variables of $\mathfrak{p}_+ \times M$ whose kernel determines all the modules with highest weights in $\Lambda_a \cap \Omega$. This result is formulated in a more precise way and given as Theorem 9.2. Since the number of cones is finite we obtain a finite number of systems which determine all the modules arising in the harmonic setting.

In section ten we extend considerations from the harmonic setting to all classical groups. The section includes explicit forms of the systems $\mathcal{D}_a$ and $\mathcal{F}_a$ for the relevant vertices. In most cases the results take the same overall form. Let $x = \begin{pmatrix} A & B \\ C & 0 \end{pmatrix}$ be a block matrix where the entries of A are vector fields on $\mathfrak{p}_+$ and the entries of B and C are certain vector fields on M. Then the systems $\mathcal{F}_a$ have bases given by the determinants or Pfaffians of matrices like x.

In the final two sections we specialize our results to two well-known sets of examples; the ladder representations and the Wallach representations. The harmonic setting can be regarded as an analysis of the decomposition of the k-fold tensor products of the oscillator representation. The components occurring in this decomposition when k equals one are called the ladder representations. In section eleven we apply the theory here to obtain eight systems of differential operators which determine through their solutions all of the ladder representations.

Section twelve includes a description of the $K_{\mathbb{C}}$-orbits which arise as the image of the map θ mentioned above. Following this we look at the Wallach representations as a special case of Theorem (7.7) which realizes modules as subspaces of coordinate rings. For these representations the corresponding harmonic polynomials span a one dimensional $\mathfrak{k}$-module and so the resulting spaces can be identified with the full coordinate ring. This connection between the Wallach representations and the coordinate rings of $K_{\mathbb{C}}$-orbits was described some years ago by Harris and Jakobsen [19]. Certainly Theorem 7.7 can be viewed as a generalization of their work.

Underlying much of this memoir is the bilinear pairing on spaces of polynomials given by polynomial differential operators. The way in which this pairing meshes with the representation theory is fundamental to all that we do. This pairing is of course well-known in representation theory and has played essential roles in work by Harris and Jakobsen [20], [28] and Kostant [33] as well as many others. In particular the reader should note that Theorem 2.9 is our version of this interplay between pairings and representations.

Other results treated here are related to work of Kostant [34]. Suppose $\mathfrak{g}$ acts on the sections of a line bundle by first order differential operators and α is a differential operator but not necessarily $\mathfrak{g}$-invariant. Kostant determines when the kernel of α is a $\mathfrak{g}$-invariant subspace. His theory is that of quasi-invariant differential operators and is closely related to our approach when our spaces of polynomials are scalar valued and when the representation of K on $\mathcal{D}_L$ is one-dimensional. We have not pursued this theme of quasi-invariance here. It is likely that aspects of what we accomplish here could be presented in terms of some notion of quasi-invariant systems of differential operators.

Representation theory is rich in structure and theorems as well as notation. With this in mind we include an index of notation following section twelve.

This project began when Davidson and Stanke asked Enright an algebraic question about the radical of a Verma module some two years ago. What has developed has gone well beyond the solution to that algebraic question and its immediate implications. The three of us have varied backgrounds and expertise in representation theory which has made this a challenging collaboration. Each of us has been broadened by the project. We hope that the analysts amongst us have kept the algebraist honest in offering a readable presentation of our ideas and vice versa.

The authors would like to thank the NSF for its support and in particular Professor Hugo Rossi for organizing workshops at the University of Utah, Salt Lake City. The authors have had distant home institutions and most of our efforts have been coordinated during these summer workshops.

Part of the research by the first named author was performed as a Fulbright fellow and research associate at the University of Iceland. He wishes to thank the Council for the International Exchange of Scholars and the Science Institute of the University of Iceland, Reykjavik.

The third author gratefully acknowledges Baylor University for partially supporting this research through University Research and Summer Sabbatical grants.

The entire manuscript was typed (and in several cases of major revisions retyped) by Neola Crimmins. Special thanks to her for her great skill in this work as well as unusual patience with the authors.

§1. Vector Bundles and Algebraic Conventions

In this section we set down the notation and define the representations which will be the focus of the rest of the article. All the results of this section are known to experts. We offer only sufficient detail to make our formulations precise.

Let G be a connected simple Lie group with faithful finite dimensional representation and let K be a maximal compact subgroup of G. Denote by $\mathfrak{g}_0$ and $\mathfrak{k}_0$ the (real) Lie algebras of G and K, respectively. Choose a Cartan subalgebra $\mathfrak{h}_0$ of $\mathfrak{k}_0$. To designate complexifications we merely omit the subscript 0. This gives complex Lie algebras $\mathfrak{g}$, $\mathfrak{k}$ and $\mathfrak{h}$.

Throughout this article we suppose G/K is a Hermitian symmetric space. Let θ denote the Cartan involution of $\mathfrak{g}$ and let $\mathfrak{g} = \mathfrak{k} \oplus \mathfrak{p}$ and $\mathfrak{g}_0 = \mathfrak{k}_0 \oplus \mathfrak{p}_0$ be the corresponding Cartan decompositions of $\mathfrak{g}$ and $\mathfrak{g}_0$. Our standing hypotheses imply that $\mathfrak{h}$ is a Cartan subalgebra of $\mathfrak{g}$ and $\mathfrak{p}$ is the unique sum of two $\mathfrak{k}$-submodules, say $\mathfrak{p} = \mathfrak{p}_+ \oplus \mathfrak{p}_-$. Moreover both $\mathfrak{p}_+$ and $\mathfrak{p}_-$ are abelian subalgebras of $\mathfrak{g}$.

Let Δ denote the set of roots of the pair $(\mathfrak{g},\mathfrak{h})$ and let Δ_c denote the subset of roots of $(\mathfrak{k},\mathfrak{h})$. Put $\Delta_n = \Delta\backslash\Delta_c$ and choose a positive system Δ^+ for Δ with $\mathfrak{p}_+$ spanned by the root spaces for elements in $\Delta_n^+ = \Delta_n \cap \Delta^+$. Similarly, put $\Delta_c^+ = \Delta_c \cap \Delta^+$. Let $\mathcal{W}$ denote the Weyl group of Δ and let $\mathcal{W}_c$ denote the subgroup generated by the reflections s_α with $\alpha \in \Delta_c$. Then $\mathcal{W}_c$ is isomorphic to the Weyl group of Δ_c. As usual we let ρ equal half the sum of the positive roots. Let $\mathcal{L}^+$ denote the Δ_c^+-dominant integral elements in $\mathfrak{h}^*$. Let $\leq$ denote the Bruhat order on $\mathfrak{h}^*$.

Our assumption that G admits a faithful finite dimensional representation implies there is a complex group $G_{\mathbb{C}}$ with Lie algebra $\mathfrak{g}$ and an imbedding of G in $G_{\mathbb{C}}$ as the analytic subgroup corresponding to the subalgebra $\mathfrak{g}_0$ of $\mathfrak{g}$. We now describe the complex structure on G/K following the approach of Harish-Chandra [18]. For more detail the reader should also consult [21]. Let $K_{\mathbb{C}}$ (resp. P_+, P_-) be the complex analytic subgroup of $G_{\mathbb{C}}$ corresponding to Lie algebra $\mathfrak{k}$ (resp. $\mathfrak{p}_+$, $\mathfrak{p}_-$). Set U equal to the semidirect product $K_{\mathbb{C}}P_-$. Then U is a parabolic subgroup of $G_{\mathbb{C}}$ and $U \cap G = K$. So G/K imbeds in $G_{\mathbb{C}}/U$ and dimension considerations imply that the image is an open G-orbit. In turn G/K inherits a complex structure where the natural action of G is by biholomorphic maps.

Next, we consider G-equivariant vector bundles on G/K and the associated representations of G. Let (π, E) be a finite dimensional holomorphic representation of $K_{\mathbb{C}}$ which we extend to a representation of U by letting P_- act by the identity. Associate to (π, E) a holomorphic vector bundle $\mathfrak{E} \to G/K$ in the usual way. In particular, $\mathfrak{E}$ is the set of classes in $G \times E$ under the equivalence relation $(gk, e) \equiv (g, k \cdot e)$, $g \in G$, $k \in K$ and $e \in E$ and the projection of $\mathfrak{E}$ onto G/K is induced by the projection π_1 onto the first coordinate. Let π_2 denote the projection of $G \times E$ onto E. Let Γ denote the space of holomorphic sections for $\mathfrak{E}$ and define an action $t_1(\cdot)$ of G on Γ by:

$$(1.1) \qquad t_1(g)\gamma(z) = g(\gamma(g^{-1}z)), \; g \in G, \; z \in G/K.$$

Here both G-actions on G/K as well as $\mathfrak{E}$ are induced by left multiplication of G on G.

The G-homogeneous bundle $\mathfrak{E}$ is trivial and so Γ is isomorphic to the E-valued holomorphic functions on G/K which we denote by $\mathcal{O}_E$. We now make this isomorphism explicit. For $g \in G$ write $g = p_+(g)k(g)p_-(g)$ with $p_+(g) \in P_+$, $k(g) \in K_{\mathbb{C}}$ and $p_-(g) \in P_-$. Such an expression is unique. For $\varphi \in \Gamma$ we may regard φ as a function on G with values in E transforming by $\varphi(gk) = \pi(k^{-1})\varphi(g)$. With this convention the action defined in (1.1) becomes left translation. Now define a function $\underline{\varphi}$ on G by $\underline{\varphi}(g) = \pi(k(g))\varphi(g)$. Since $k(gk_1) = k(g)k_1$, for $g \in G$ and $k_1 \in K$, $\underline{\varphi}$ is constant on right K-orbits; and so induces a function $\varphi^{\vee}$ on G/K. The assignment $\varphi \to \varphi^{\vee}$ is the isomorphism of Γ onto $\mathcal{O}_E$. The action t_1 of G on Γ can then be transferred to an action t_2 on $\mathcal{O}_E$. We obtain the formula:

$$(1.2) \qquad t_2(g)\psi(xK) = \pi(k(x)k(g^{-1}x)^{-1})\psi(g^{-1}xK), \quad g,x \in G.$$

Our immediate goal is to describe the infinitesimal action of the Lie algebra $\mathfrak{g}$ and its universal enveloping algebra $U(\mathfrak{g})$ on the set of K-finite vectors in this representation t_2 on $\mathcal{O}_E$. A neighborhood N of zero in $\mathfrak{p}_+$ maps biholomorphically onto a neighborhood of eK in G/K by the exponential map, exp. Any holomorphic function on G/K is determined by its restriction to N. Moreover in $\mathfrak{p}_+$ coordinates the formula (1.2) simplifies somewhat to:

$$(1.3) \qquad t_2(g)\psi(xK) = \pi(k(g^{-1}x)^{-1})\psi(g^{-1}xK), \quad \text{for } x = \exp z,\ z \in \mathfrak{p}_+.$$

Also, since $\mathfrak{p}_+$ is $\mathrm{ad}(\mathfrak{k})$-stable,

(1.4) $$t_2(k)\psi(xK) = \pi(k)\psi(k^{-1}xK) = \pi(k)\psi(\exp(\mathrm{Ad}(k^{-1})z)K),$$
$$\text{for } x = \exp z,\ z \in \mathfrak{p}_+.$$

For vector spaces V and W, let P(V,W) be the algebra of W-valued polynomials on V. When W is one dimensional we write P(V) in place of P(V,W). Let S(V) denote the symmetric algebra of V. If V is finite dimensional and V and V' are paired by a nondegenerate bilinear form φ then φ induces a natural isomorphism: $P(V,W) \simeq S(V') \otimes W$.

1.5 CONVENTION.

(i) For any finite dimensional $\mathfrak{k}$-module E let $\mathbb{E}$ denote the space $P(\mathfrak{p}_+,E)$ of E-valued polynomials on $\mathfrak{p}_+$.

(ii) The Killing form gives a nondegenerate pairing of $\mathfrak{p}_+$ and $\mathfrak{p}_-$ and so we have canonical isomorphisms:

$$\mathbb{E} \simeq S(\mathfrak{p}_-) \otimes E.$$

Let t denote the infinitesimal action of $\mathfrak{g}$ (or $U(\mathfrak{g})$) induced by t_2 on the K-finite vectors in $\mathcal{O}_E$.

1.6 LEMMA.

(i) The K-finite vectors in $\mathcal{O}_E$ (under the action t_2) are precisely the E-valued polynomials on $\mathfrak{p}_+$, $\mathbb{E}$.

(ii) The t-action of $\mathfrak{k}$ is just the tensor product action on $S(\mathfrak{p}_-) \otimes E$.

(iii) For $X \in \mathfrak{p}_+$, $t(X)$ act by minus differentiation in the X-direction (i.e. by directional derivatives).

(iv) The $\mathfrak{k}$-module $1 \otimes E$ is cocyclic in $S(\mathfrak{p}_-) \otimes E$ (i.e. $1 \otimes E$ is contained in every nonzero $\mathfrak{g}$-submodule).

PROOF. The elements of $\mathcal{O}_E$ can be identified with certain E-valued power series on $\mathfrak{p}_+$ by using $\mathfrak{p}_+$ coordinates on G/K via exp. Choose an element Z_0 in the center of $\mathfrak{k}$ such that $ad(Z_0)$ equals the identity on $\mathfrak{p}_+$. Our hypotheses on (G,K) imply Z_0 exists (and in fact it is unique). Then $ad(Z_0)$ acts by $-m$ times the identity on the homogeneous polynomials of degree m. In turn the only $\mathfrak{k}$-finite power series with values in E must be polynomials. Conversely, exp is a diffeomorphism of $\mathfrak{p}_+$ onto G/K; and so $\mathbb{E}$ is contained in $\mathcal{O}_E$. This proves (i). Then (ii) and (iii) follow from formulas (1.4) and (1.3), respectively. Finally (iii) implies (iv).

The $\mathfrak{g}$-module defined by $\mathfrak{t}$ is equivalent to a certain coinduced module. We now describe this coinduced $\mathfrak{g}$-module. Let $\mathfrak{q}$ (resp. $\bar{\mathfrak{q}}$) denote the parabolic subalgebra $\mathfrak{k} \oplus \mathfrak{p}_+$ (resp. $\mathfrak{k} \oplus \mathfrak{p}_-$). Define the $\mathfrak{g}$-module coinduced from the $\mathfrak{q}$-modules A as the space $L = \mathrm{Hom}_{U(\mathfrak{q})}(U(\mathfrak{g}),A)$ where the linear map f is an element of L if $f(ab) = af(b)$ for $a \in U(q)$ and $b \in U(\mathfrak{g})$ with the $\mathfrak{g}$-module action given by the right translation: $(x \cdot f)(a) = f(ax)$, $a, x \in U(\mathfrak{g})$.

Coinduced modules are duals to induced modules in the following precise sense. For more detail the reader should consult [11]. Let $u \mapsto u^T$ denote the principal antiautomorphism of $U(\mathfrak{g})$. Let A^* denote the algebraic dual of A. Then, for $\psi \in \mathrm{Hom}_{U(\mathfrak{q})}(U(\mathfrak{g}),A^*)$, $u \in U(\mathfrak{g})$ and $a \in A$, define a bilinear pairing of $\mathrm{Hom}_{U(\mathfrak{q})}(U(\mathfrak{g}),A^*)$ and $U(\mathfrak{g}) \otimes_{U(\mathfrak{q})} A$ by the formula

$$\langle \psi, u \otimes a \rangle = \psi(u^T)(a). \tag{1.7}$$

This pairing induces an isomorphism of $\mathrm{Hom}_{U(\mathfrak{q})}(U(\mathfrak{g}), A^*)$ onto the algebraic dual to the induced module $U(\mathfrak{g}) \underset{U(\mathfrak{q})}{\otimes} A$.

The pairing above is the standard one. However, for our purposes we shall need conjugate linear pairings instead. Let σ be the conjugate linear antiautomorphism of $U(\mathfrak{g})$ induced by the map $X \mapsto -X$ on the real form $\mathfrak{g}_0$ of $\mathfrak{g}$. Since $\mathfrak{h} \subset \mathfrak{k}$ and all roots are pure imaginary on $\mathfrak{h} \cap \mathfrak{k}_0$, we find that σ fixes the co-roots in $\mathfrak{h}$ and for all roots γ in Δ, $\sigma\mathfrak{g}_\gamma = \mathfrak{g}_{-\gamma}$. In particular, $\sigma\mathfrak{p}_+ = \mathfrak{p}_-$ and $\sigma^2 = 1$.

Let $\varphi = \langle \cdot,\cdot \rangle$ denote a nondegenerate sesquilinear pairing of two vector spaces B and C. Suppose B is a $\mathfrak{g}$-module with action denoted by $r(\cdot)$. Then, for $X \in \mathfrak{g}$, $X \to$ adjoint of $r(\sigma X)$ defines a $\mathfrak{g}$-module action on C which we call the conjugate dual representation to B.

If B and C are $\mathfrak{g}$-modules, then the pairing φ is called $\mathfrak{g}$ (resp. $\mathfrak{k}$)-invariant if $\langle xb, c\rangle = \langle b, \sigma(x)c\rangle$ for all $b \in B$, $c \in C$, and $x \in U(\mathfrak{g})$ (resp. $U(\mathfrak{k})$).

1.8 LEMMA. Suppose E is a finite dimensional $\mathfrak{k}$-module with $\mathfrak{k}$-invariant Hermitian form $\langle \cdot,\cdot \rangle_E$. View E separately as both a $\mathfrak{q}$ and $\bar{\mathfrak{q}}$ module by letting $\mathfrak{p}_+$ and $\mathfrak{p}_-$ act by zero in the two cases. With this convention we have:

(i) The $\mathfrak{k}$-finite vectors in $\mathrm{Hom}_{U(\bar{\mathfrak{q}})}(U(\mathfrak{g}), E)$ and $U(\mathfrak{g}) \underset{U(\mathfrak{q})}{\otimes} E$ are both canonically isomorphic as $\mathfrak{k}$-modules to $S(\mathfrak{p}_-) \otimes E$.

(ii) The two $\mathfrak{g}$-modules in (i) are the $\mathfrak{k}$-finite conjugate dual modules of each other.

PROOF. Let X denote the $\mathfrak{g}$-module of $\mathfrak{k}$-finite vectors in $\mathrm{Hom}_{U(\bar{\mathfrak{q}})}(U(\mathfrak{g}),E)$. By the Poincaré-Witt-Birkhoff theorem, $U(\mathfrak{g}) \otimes_{U(\mathfrak{q})} E \simeq S(\mathfrak{p}_-) \otimes E$ and X is $\mathfrak{k}$-isomorphic to the $\mathfrak{k}$-finite vectors in $\mathrm{Hom}_{\mathbb{C}}(S(\mathfrak{p}_+),E)$. Since $\mathfrak{p}_+$ and $\mathfrak{p}_-$ are paired by the Killing from this gives (i).

The proof of (ii) follows the proof in the bilinear case [cf. 11, §5.5] with the bilinear pairing (1.7) replaced by the conjugate linear pairing: for $\psi \in X$, $u \in U(\mathfrak{g})$ and $e \in E$ define

$$\langle \psi, u \otimes e \rangle = \langle \psi(\sigma u), e \rangle_E . \tag{1.9}$$

The geometric and algebraic representations are related by the following lemma.

1.10 LEMMA. The representation $\mathfrak{t}$ of $\mathfrak{g}$ on $\mathbb{E}$ is equivalent to the $\mathfrak{g}$-module of $\mathfrak{k}$-finite vectors in the coinduced module $\mathrm{Hom}_{U(\bar{\mathfrak{q}})}(U(\mathfrak{g}),E)$.

PROOF. Let X denote the $\mathfrak{k}$-finite vectors in $\mathrm{Hom}_{U(\bar{\mathfrak{q}})}(U(\mathfrak{g}),E)$ and Y denote the $\mathfrak{g}$-module $\mathbb{E}$ with action given by $\mathfrak{t}$. From (1.6) and (1.8), X and Y are $\mathfrak{k}$-isomorphic. Let Z be any $\mathfrak{k}$-finite $\mathfrak{g}$-module and π a $\bar{\mathfrak{q}}$-module projection $\pi\colon Z \to E$. Then, for $z \in Z$, $\varphi_z(x) = \pi(x \cdot z)$ defines an element of X. Moreover, if E is cocyclic in Z then the assignment $z \mapsto \varphi_z$ is an injective $\mathfrak{g}$-module map. Applying this with $Z = Y$ and noting that X and Y are $\mathfrak{k}$-isomorphic with finite multiplicities implies that the map $z \mapsto \varphi_z$ is an isomorphism of Y onto X.

§2. Conjugate Pairings and Reproducing Kernels

For spaces of polynomials, linear and conjugate linear pairings via polynomial differential operators are well known. Within representation theory these pairings have played a role in several contexts (cf. [33], [20] and [28]). For any representation on the space of polynomials $P(\mathfrak{p}_+, E)$, this pairing will associate the conjugate dual representation on the same space of polynomials. The interplay between polynomials and polynomial differential operators and these two representations is fundamental to all that follows. The results of this section are offered in more detail but are very similar to the use of pairings found in [20].

We keep in force the notation from section one and further assume that the $\mathfrak{k}$-module E is irreducible. Let L denote the unique simple submodule of the representation $\mathfrak{t}$ on $P(\mathfrak{p}_+, E)$ (cf. (1.6)(iv)). In this section we characterize L as the kernel of a system of constant coefficient differential operators acting on these E-valued polynomials.

Let V and W be finite dimensional complex vector spaces with inner products $(\,|\,)_V$ and $(\,|\,)_W$, respectively. We omit the subscripts when no confusion can arise. Let dT denote Lebesgue measure on V and $d\mu$ the Gaussian measure $e^{-(T|T)}dT$ with dT normalized so that $\int_V d\mu = 1$. Now define an inner product on $P(V, W)$ by:

$$(F|G) = \int_V (F(T)|G(T))_W \, d\mu. \tag{2.1}$$

Next we define the conjugate linear map from $P(V,W)$ to a space of differential operators. Fix any basis $\{e_1,\ldots,e_m\}$ of V and let $\alpha = (\alpha_1,\ldots,\alpha_m) \in \mathbb{N}^m$ be a multi-index. Define the monomial E^α in $P(V)$ by $E^\alpha(T) = \prod_i (T|e_i)^{\alpha_i}$. For any $w \in W$ we regard $E^\alpha w$ as a monomial in $P(V,W)$. For any $R \in V$ let D_R denote the (nonnormalized) directional derivative along R defined by $D_RF(T) = \lim_{t\to 0} \frac{F(T+tR)-F(T)}{t}$. Then define ∂^α as the product $\prod D_{e_i}^{\alpha_i}$. For $G \in P(V,W)$ we have a unique expansion $G = \Sigma_\alpha E^\alpha w_\alpha$ with $w_\alpha \in W$. Define the differential operator $G(\partial)$ to be the linear map from $P(V,W)$ to $P(V)$ given by the formula:

$$G(\partial)F = \sum_\alpha (\partial^\alpha F|w_\alpha)_W \,. \tag{2.2}$$

Note that $G \mapsto G(\partial)$ is conjugate linear and in the case $P = \Sigma_\alpha a_\alpha E^\alpha$, $a_\alpha \in \mathbb{C}$,

$$P(\partial) = \sum_\alpha \bar{a}_\alpha \partial^\alpha \,. \tag{2.3}$$

The space of polynomials $P(V,W)$ has a reproducing kernel $(S,T) \mapsto e^{(S|T)}$. So we obtain the formula: for $F \in P(V,W)$

$$F(S) = \int_V F(T)e^{(S|T)}d\mu(T)\,. \tag{2.4}$$

Now reversing the orders of integration and differentation we obtain the following formula for the action of directional derivatives D_R, for $R \in V$

$$D_RF(S) = \int_V F(T)e^{(S|T)}(R|T)d\mu(T)\,. \tag{2.5}$$

Combining this formula with (2.2) and (2.3) we obtain

$$G(\partial)F(S) = \int_V (F(T)|G(T))e^{(S|T)}d\mu(T) \tag{2.6}$$

and

(2.7) $$P(\partial)F(S) = \int_V F(T)\,\overline{P(T)}\,e^{(S|T)}d\mu(T).$$

Note that these formulas imply that the operators $G(\partial)$ and $P(\partial)$ are independent of the basis chosen.

These formulas allow us to express the inner product on $P(V,W)$ in terms of differential operators.

2.8 LEMMA. Let $F, G \in P(V,W)$ and $P \in P(V)$. Then

(i) $(F|G) = G(\partial)F(0)$ and

(ii) $(F|PG) = (P(\partial)F|G) = (G(\partial)F|P)$.

PROOF. The first identity follows by letting S equal zero in (2.6). The second follows from the first by observing that all three expressions equal $(PG)(\partial)F(0)$.

We now return to representation theory and the setting of section one. With $(\mathfrak{t}, \mathbb{E})$ as in (1.5), assume that E is an irreducible $\mathfrak{k}$-module with $\mathfrak{k}$-invariant Hermitian form $\langle\ ,\ \rangle_E$. Recall from the discussion following (1.7) the conjugate linear antiautomorphism σ of $U(\mathfrak{g})$. The space $\mathbb{E}$ is a space of E-valued polynomials and so $\mathbb{E}$ has an inner product given by (2.1) above. Using this inner product we associate to any $\mathfrak{k}$-finite linear map on $\mathbb{E}$ its adjoint. Then, for $x \in U(\mathfrak{g})$, the assignment $x \mapsto$ adjoint $\mathfrak{t}(\sigma x)$ defines another representation of $\mathfrak{g}$ on $\mathbb{E}$ called the conjugate dual to $\mathfrak{t}$. Denote this representation by $\mathfrak{t}^*$. From Lemmas 1.8 and 1.10 we know that $(\mathfrak{t}^*, \mathbb{E})$ is equivalent to the generalized Verma module representation on $S(\mathfrak{p}_-) \otimes E$. We should note in particular that this pairing of $\mathfrak{t}$ and $\mathfrak{t}^*$ carries the directional derivative action along $z \in \mathfrak{p}_+$ (i.e., the action of $\mathfrak{t}(z)$) to multiplication by σz on $\mathbb{E}$ (i.e. the action of $\mathfrak{t}^*(\sigma z)$).

Let L be the unique simple submodule of $(\mathfrak{t},\mathbb{E})$ and let J be the unique maximal submodule of $(\mathfrak{t}^*,\mathbb{E})$. Then J is the orthogonal complement of L and vice versa. Since E is finite dimensional and $S(\mathfrak{p}_-)$ is a Noetherian ring, J is finitely generated as a module over $S(\mathfrak{p}_-)$.

2.9 THEOREM. Let $m_1,\ldots,m_t$ be any set of generators for J as an $S(\mathfrak{p}_-)$-module. Then the simple submodule L is the kernel of the constant coefficient operators $m_i(\partial)$, $1 \le i \le t$.

PROOF. For $F \in P(\mathfrak{p}_+,E)$,

$$\begin{aligned} F \in L \iff & (F, p\cdot m_i) = 0 \text{ for all } p \in P(\mathfrak{p}_+,\mathbb{C}),\ 1 \le i \le t, \\ \iff & p(\partial)(m_i(\partial)F)(0) = 0 \text{ for all } p \in P(\mathfrak{p}_+,\mathbb{C}),\ 1 \le i \le t, \\ \iff & m_i(\partial)F \equiv 0 \quad \text{for } 1 \le i \le t. \end{aligned}$$

The next section will give a simple description of the generators of J when L is a unitarizable module.

§3. $\mathfrak{k}$-Irreducibility of the System of Differential Operators

In this section we sharpen the results from section two in the cases when L is unitarizable. Let notation be as in earlier sections. In particular L and J are respectively the irreducible and maximal submodules of $(\mathfrak{t}, \mathbb{E})$ and $(\mathfrak{t}^*, \mathbb{E})$. For any $\mathfrak{p}_+$-module A, let $H^i(\mathfrak{p}_+, A)$ denote the ith Lie algebra cohomology group.

3.1 THEOREM. Suppose L is unitarizable and $J \neq 0$. Then

(i) $H^1(\mathfrak{p}_+, L)$ is an irreducible $\mathfrak{k}$-module.

(ii) J is generated over $S(\mathfrak{p}_-)$ by an irreducible $\mathfrak{k}$-submodule J^o isomorphic to $H^1(\mathfrak{p}_+, L)$.

(iii) L is determined as the kernel of the system of differential operators $m_i(\partial)$ where the span of $\{m_i\}$ is J^o.

Note: Formulas for the highest weight of J^o are given in (3.8) as well as (6.8). The operators $m_i(\partial)$ in (iii) are homogeneous and all of the same degree. This degree is given in (3.10).

3.2 DEFINITION. When L is unitarizable and $\{m_1, \ldots, m_r\}$ is a (linear) basis of J^o we let $\mathcal{D}_L$ denote the system of differential operators $m_j(\partial)$, $1 \leq j \leq r$. If L has the highest weight $\lambda - \rho$ we also write $\mathcal{D}_\lambda$ for $\mathcal{D}_L$.

The proof of (3.1) will be completed for the classical groups in this section and for the exceptional groups in the next.

For $\xi \in \mathcal{L}^+$, let $F = F(\xi)$ denote the irreducible $\mathfrak{k}$-module with highest weight ξ. Consider F as a $\mathfrak{q}$-module by letting $\mathfrak{p}_+$ act by zero and define the associated generalized Verma modules by:

$$N(\xi + \rho) = U(\mathfrak{g}) \underset{U(\mathfrak{q})}{\otimes} F. \tag{3.3}$$

If the $\mathfrak{k}$-module E used above has highest weight $\lambda - \rho$, then L is isomorphic to the unique irreducible quotient $L(\lambda)$ of $N(\lambda)$.

Let $\mathcal{O} = \mathcal{O}(\mathfrak{g}, \mathfrak{q})$ denote the category of $\mathfrak{g}$-modules which are finitely generated, $U(\mathfrak{q})$-locally finite and $U(\mathfrak{k})$-completely reducible. This category is by now well understood. For detailed expositions the reader should consult [30] or [16]. Let $\mathrm{Ext}^i(A,B)$ denote the derived functors of Hom calculated in this category. Connections with the cohomology spaces $H^i(\mathfrak{p}_+, \ )$ can be found in [41].

To prove (3.1) it will be convenient to express the assertion in terms of Ext groups.

3.4 DEFINITION. Let $\nu \in \mathcal{L}^+$. Then we say the pair (λ, ν) satisfies condition (3.4) if

(i) $\mathrm{Hom}_{\mathfrak{g}}(N(\nu), N(\lambda)) \neq 0$ and

(ii) $\mathrm{Ext}^1(N(\mu), L(\lambda)) = 0$ for all $\mu \in \mathcal{L}^+$ with $\mu \leq \lambda$ and $\mu \neq \nu$.

3.5 LEMMA. Set $L = L(\lambda)$. For any $\nu \in \mathcal{L}^+$, (3.4) holds for the pair (λ, ν) if and only if L is isomorphic to the quotient of $N(\lambda)$ by a nonzero image of $N(\nu)$. Moreover, when these equivalent conditions hold, $\mathrm{Ext}^1(N(\nu), L) \simeq \mathbb{C}$.

PROOF. It is sufficient to consider the case where $N(\lambda)$ is reducible. Suppose (3.4) does not hold. Then for some $\mu \neq \nu$, $\mu \leq \lambda$, $\mathrm{Ext}^1(N(\mu),L) \neq 0$. Let $K = L(\mu)$. By a result of Barbasch [5] this implies $\mathrm{Ext}^1(K,L) \neq 0$. Taking contravariant duals we obtain a module X and nonsplit short exact sequence $0 \to K \to X \to L \to 0$. Now $\lambda - \rho$ is a highest weight for X and so X is a quotient of $N(\lambda)$. Thus there exists a $\mathfrak{g}$-submodule J^1 of the maximal submodule J in $N(\lambda)$ with $X \simeq N(\lambda)/J^1$ and $J/J^1 \simeq K$. But any image of $N(\nu)$ has only $L(\nu)$ as a simple quotient; and so, J is not equal to the image of $N(\nu)$.

Conversely, suppose J is not equal to the image I of $N(\nu)$. If $I = 0$ then (3.4)(i) does not hold and so we may assume $I \neq 0$. Choose a maximal submodule J^1, $I \subset J^1 \subset J$. Then $K = J/J^1 \simeq L(\mu)$ for some μ. Since $N(\lambda)$ is multiplicity free [6] or [16] $\mu \neq \nu$. Thus $N(\lambda)/J^1$ gives a nontrivial element in $\mathrm{Ext}^1(L,K) \simeq \mathrm{Ext}^1(K,L) \simeq \mathrm{Ext}^1(N(\mu),L)$. The last isomorphism again coming from [5]. So (3.4(ii)) does not hold.

In the cases when (3.4) holds the first argument above shows that any element of $\mathrm{Ext}^1(L,L(\nu))$ is represented by a module X which is a quotient of $N(\lambda)$. But J has a unique maximal submodule J^1 in this case and so the only possible X is $N(\lambda)/J^1$. This gives $\mathrm{Ext}^1(N(\nu),L) \simeq \mathrm{Ext}^1(L,L(\nu)) \simeq \mathbb{C}$ completing the proof.

To apply the previous lemma to our setting we need information on the various Ext^1 groups or equivalently information on $H^1(\mathfrak{p}_+,L)$. We now briefly summarize what is known for G classical.

When the root system Δ contains only one root length we regard all roots as short by convention.

3.6 DEFINITION.

(a) Let $\lambda \in \mathfrak{h}^*$. Then define a subgroup $\mathcal{W}_\lambda$ of $\mathcal{W}$ to be the subgroup generated by the identity and the refections s_α which satisfy

(i) $\alpha \in \Delta_n^+$ and $(\lambda, \alpha^\vee) \in \mathbb{N}^*$,

(ii) if $\beta \in \Delta$ and $(\lambda, \beta) = 0$ then $(\alpha, \beta) = 0$,

(iii) if there exists $\beta \in \Delta$ with β long and $(\lambda, \beta) = 0$ then α is short.

Let Δ_λ be the subset of Δ of elements β with $s\beta \in \mathcal{W}_\lambda$ and let $\Delta_{\lambda,c} = \Delta_c \cap \Delta_\lambda$, $\Delta_\lambda^+ = \Delta_\lambda \cap \Delta^+$ and $\Delta_{\lambda,c}^+ = \Delta_{\lambda,c} \cap \Delta^+$. Then Δ_λ is an abstract root system [12]. Let ℓ_λ denote the length function on $\mathcal{W}_\lambda$ determined by Δ_λ^+.

(b) Put $\mathcal{W}_{\lambda,c} = \mathcal{W}_\lambda \cap \mathcal{W}_c$ and for $i \in \mathbb{N}$, set $\mathcal{W}_\lambda^{c,i} = \{\omega \in \mathcal{W}_\lambda \mid \omega\rho$ is $\Delta_{\lambda,c}^+$-dominant and $\ell_\lambda(\omega) = i\}$ and $\mathcal{W}_\lambda^c = \cup \mathcal{W}_\lambda^{c,i}$. Then $\mathcal{W}_\lambda = \mathcal{W}_{\lambda,c} \cdot \mathcal{W}_\lambda^c$.

(c) Let $(\mathfrak{g}_\lambda, \mathfrak{k}_\lambda)$ be the pair of Lie algebras each with Cartan subalgebra $\mathfrak{h}$ and root systems Δ_λ and $\Delta_{\lambda,c}$, respectively. These are the complexified Lie alegbras of a Hermitian symmetric pair (G_λ, K_λ).

(d) Let $\xi \in \mathfrak{h}^*$ and suppose $\langle \xi, \alpha \rangle$ is real for all $\alpha \in \Delta_c$. Then the $\mathcal{W}_c$-orbit of ξ has a unique Δ_c^+-dominant element. Let $\bar{\xi}$ denote this element.

3.7 THEOREM. [12] Suppose G is classical and L is unitarizable with highest weight $\lambda - \rho$. Then for $i \in \mathbb{N}$,

$$H^i(\mathfrak{p}_+, L) \simeq \bigoplus_{\omega \in \mathcal{W}_\lambda^{c,i}} F(\overline{\omega\lambda} - \rho).$$

3.8 COROLLARY. Let γ_0 be the unique noncompact simple root in the positive root system Δ_λ^+ and let $\lambda_0 = \overline{s_{\gamma_0}\lambda}$. Then

$$H^1(\mathfrak{p}_+, L) \simeq F(\lambda_0 - \rho) \quad \text{and} \quad \mathrm{Hom}_{\mathfrak{g}}(N(\lambda_0), N(\lambda)) \neq 0.$$

The first assertion follows directly from (3.7). For the second we combine Proposition 3.2 in [12] with the formulas for Hom given as Theorem 8.6 in [16].

Proof of (3.1) for G classical. With notation as in (3.8), for $\mu \in \mathcal{L}^+$ we have isomorphisms (cf. §7 [41]):

$$\text{(3.9)} \qquad \mathrm{Ext}^i(N(\mu), L) \simeq \mathrm{Hom}_{\mathfrak{k}}(F(\mu - \rho), H^i(\mathfrak{p}_+, L)).$$

Combining this with (3.8) we conclude that the pair (λ, λ_0) satisfies (3.4). But then (3.5) asserts that J is the image of the generalized Verma module $N(\lambda_0)$. This proves (ii). Parts (i) and (iii) follow directly from (3.8) and (2.9) respectively.

3.10 REMARKS.

(i) The root γ_0 can be defined as the minimal element in the set $\{\alpha \in \Delta_n^+ \mid (\lambda, \alpha^\vee) \in \mathbb{N}^*$, α is orthogonal to any β with $(\lambda, \beta) = 0$ and, if $(\lambda, \beta) = 0$ for any long root β then α is short$\}$.

(ii) The system of differential operators $\mathcal{D}_\lambda$ is made up of homogeneous elements all of degree $(\lambda, \gamma_0^\vee)$. This integer is equal to the "level of reduction" of $N(\lambda)$ as defined in [15] (cf. section six for the definition of level of reduction).

(iii) We will give another explicit expression for $\overline{s_{\gamma_0}\lambda}$ in (6.8).

§4. $\mathfrak{p}_+$-Cohomology for the Exceptional Groups

Keep in force the notation from the previous sections and recall in particular Definition (3.6). We now prove the $\mathfrak{p}_+$-cohomology formulas for G exceptional and complete the proof of (3.1).

4.1 THEOREM. Suppose L is unitarizable with highest weight $\lambda - \rho$. Then, for $i \in \mathbb{N}$,

$$H^i(\mathfrak{p}_+, L) \simeq \bigoplus_{\omega \in \mathcal{W}_\lambda^{c,i}} F(\overline{\omega\lambda} - \rho).$$

The results in the Memoir [16] have now been extended to include the exceptional groups [17]. This extension is the starting point for the proof of (4.1). In [8] Collingwood introduced the definition of a Kostant module for a highest weight module with a regular integral parameter. We generalize his definition as follows:

4.2 DEFINITION. If L is an irreducible highest weight module with highest weight $\lambda - \rho$ we call L a Kostant module if for all $i \in \mathbb{N}$ the cohomology formulas in (4.1) hold.

For regular integral λ this definition is equivalent to Collingwood's since in this case the length function ℓ_λ will equal the restriction to $\mathcal{W}_\lambda$ of the length function on $\mathcal{W}$. For such a parameter the Δ_c^+-dominant elements in the $\mathcal{W}$-orbit correspond to the vertices of the posets given below in Figures 4.5 and 4.6 for the cases of E_6 and E_7, respectively. The points which correspond to Kostant modules are circled.

Next we turn to parameters λ which are semiregular (i.e., λ lies on exactly one root hyperplane).

4.3 PROPOSITION. Suppose λ is semiregular.

(i) For G of type E_6, L is a Kostant module.

(ii) For G of type E_7 the semiregular category containing L is equivalent to the regular integral category for the Hermitian symmetric pair (D_6, D_5) and has poset given by:

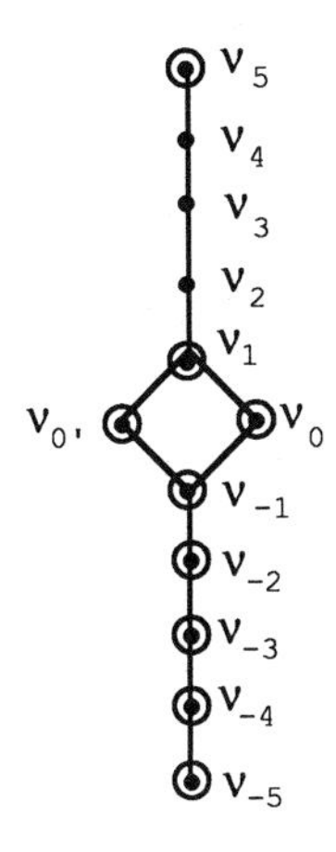

The vertices which are circled correspond to the Kostant modules. (Refer to [17] for any unexplained terms). Also, for all Kostant modules except $\lambda = \nu_1$ the set $\mathcal{W}_\lambda^{\mathfrak{c},1}$ contains one element.

PROOF. This is an immediate consequence of Lemma 4.1 and Proposition 2.3 and 3.9 in [17].

4.4 PROPOSITION. Suppose G is of type E_7. Let $\lambda - \rho \in \mathcal{L}^+$ and assume λ lies on exactly two root hyperplanes. Then L is a Kostant module.

PROOF. Any λ satisfying our hypotheses is said to have an $A_1 \times A_1$ singularity. In [17] Lemma 3.2 asserts that all categories $\mathcal{O}(\lambda)$ for these λ are equivalent and they resemble the regular integral category for $s\ell(2)$. Since G has split rank three there can be at most one reflection $s_\gamma \in \mathcal{W}_\lambda$. Then using Figure 4.6 we can verify that $\mathcal{W}_\lambda$ is trivial when $N(\lambda)$ is irreducible and $\mathcal{W}_\lambda = \{1, s_\gamma\}$ otherwise. This proves the propositon.

We now turn to the proof of (4.1). The classification of unitary highest weight modules given in [15] was established in part through a case by case analysis in sections twelve and thirteen of that article. We shall adhere to that same notation here. Any undefined terms are from those sections.

First suppose G is of type E_6 and assume L is unitarizable. For the cases Case I,p ($1 \le p \le 5$) only the end point of each line corresponds to a module where $J \neq 0$ (i.e. there is only one reduction point). For $p = 1$, 2 or 3 this point λ is regular integral and corresponds to the vertices 1, 2 and 3, respectively, in Figure 4.5. Each of these is a Kostant module. For Case I,4 λ is regular and corresponds to either vertex 4 or 5 in Figure 4.5 or λ is semiregular. For Case I,5 λ is either semiregular or regular and corresponds to vertex 7 in Figure 4.5. All of these are Kostant modules.

For each line in Case II there are two points to consider; the first reduction point ν and the "isolated" end point μ. All of the points ν are semiregular. The points μ are either semiregular or are regular integral and correspond to the vertex 12. All of these modules are Kostant modules.

The last case, Case III, includes two points. One is the trivial representation and the other is a semiregular point. Both are Kostant modules.

Now suppose G is of type E_7. Here also we have three cases. For each line in Case I,p $(1 \leq p \leq 6)$ there is only one reduction point, say λ. For p = 1, 2, 3 and 4, λ is regular integral and corresponds to the vertices 1, 2, 3 and 4, respectively, in Figure 4.6. For p = 5 we obtain either a regular integral point corresponding to the vertex 5 or 6 or a semiregular point of type ν_{-4} (cf.(4.3(ii))). For p = 6, λ is regular integral and corresponds to vertex 8 or λ is semiregular and of type ν_{-3}. In all these cases the modules are Kostant modules.

For Case II we have two types of points; the first reduction points ν and "isolated" points μ. The points ν are all semiregular and are of type ν_{-4}. The points μ have a more complicated description. For this case there is one simple compact root say δ with the highest weight not singular at δ. Put $r = (\lambda, \delta^\vee)$. Then r is an integer ≥ 2. In the subcase r = 2, λ has an $A_1 \times A_1$ singularity and we may apply (4.4). For the subcases r = 3, 4 or 5, λ is semiregular and has type ν_0, and we apply (4.3)(ii). Finally for $r \geq 6$, λ is regular integral and corresponds to vertex 16 in Figure 2. In all cases L is a Kostant module.

We now come to the last case, Case III. Here we have the trivial representation, the first reduction point ν and one other say μ. The point ν has an $A_1 \times A_1$ singularity and so we apply (4.4). The point μ is semiregular and has type v5. So all modules here are Kostant modules. This completes the proof of Theorem 4.1.

Completion of the proof of Theorem 3.1. Theorem (3.7) now holds for all G and an inspection of the cases above indicate that the sets $\mathcal{W}_\lambda^{c,1}$ all contain exactly one element, the reflection s_{γ_0} with γ_0 as in (3.10). Moreover, as in (3.8), $\mathrm{Hom}_{\mathfrak{g}}(N(\lambda_0), N(\lambda)) \neq 0$. For the regular integral points λ this follows from the extension of Theorem 8.6 in [16] to the exceptional cases given in [17]. For the semiregular points this follows directly from the analysis of these categories in [17] and for the λ with $A_1 \times A_1$ singularities it follows since the category is equivalent to $\mathcal{O}$ for sl(2). Therefore the proof for the classical G now applies in general.

Figure 4.5
(E_6, D_5)

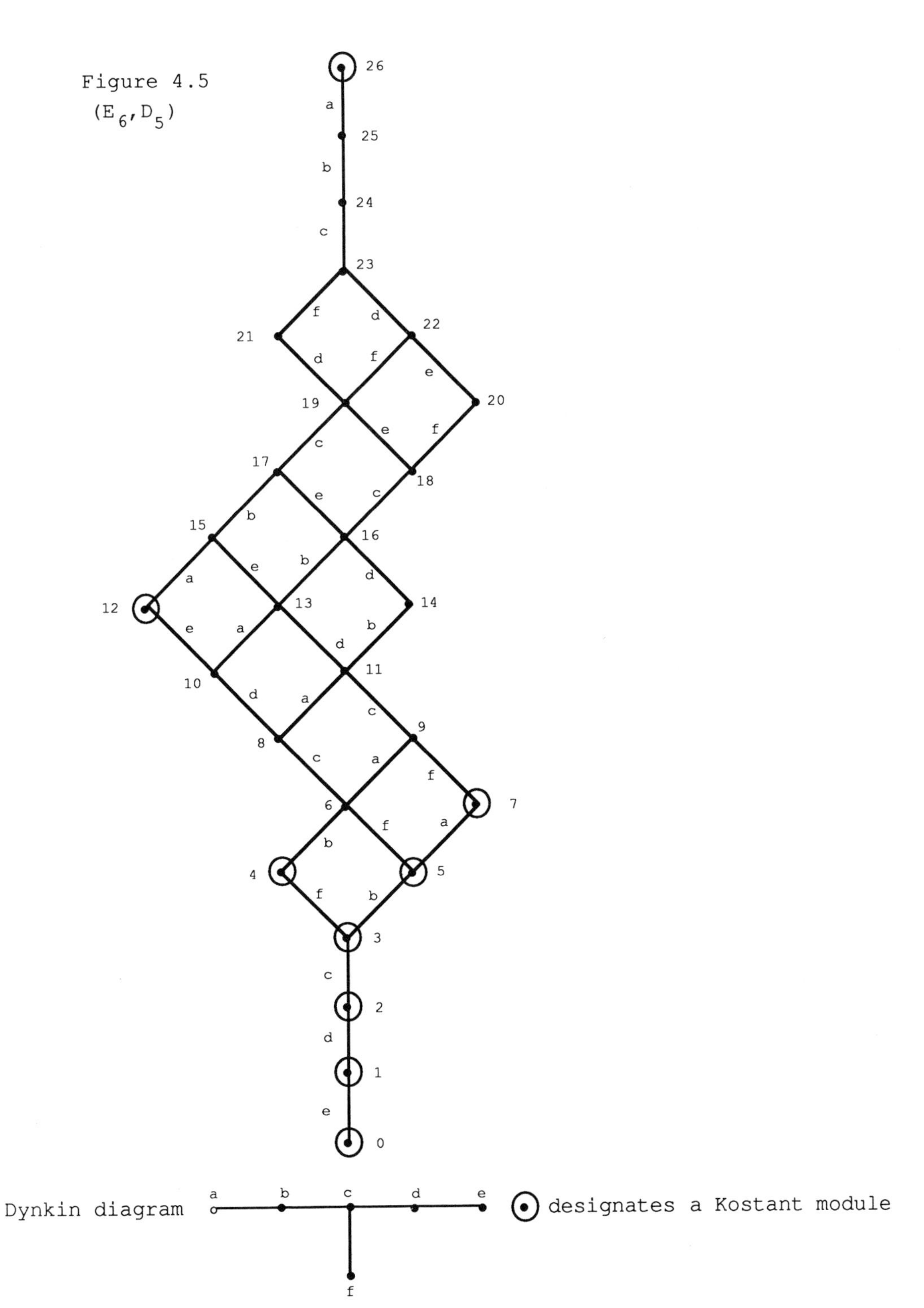

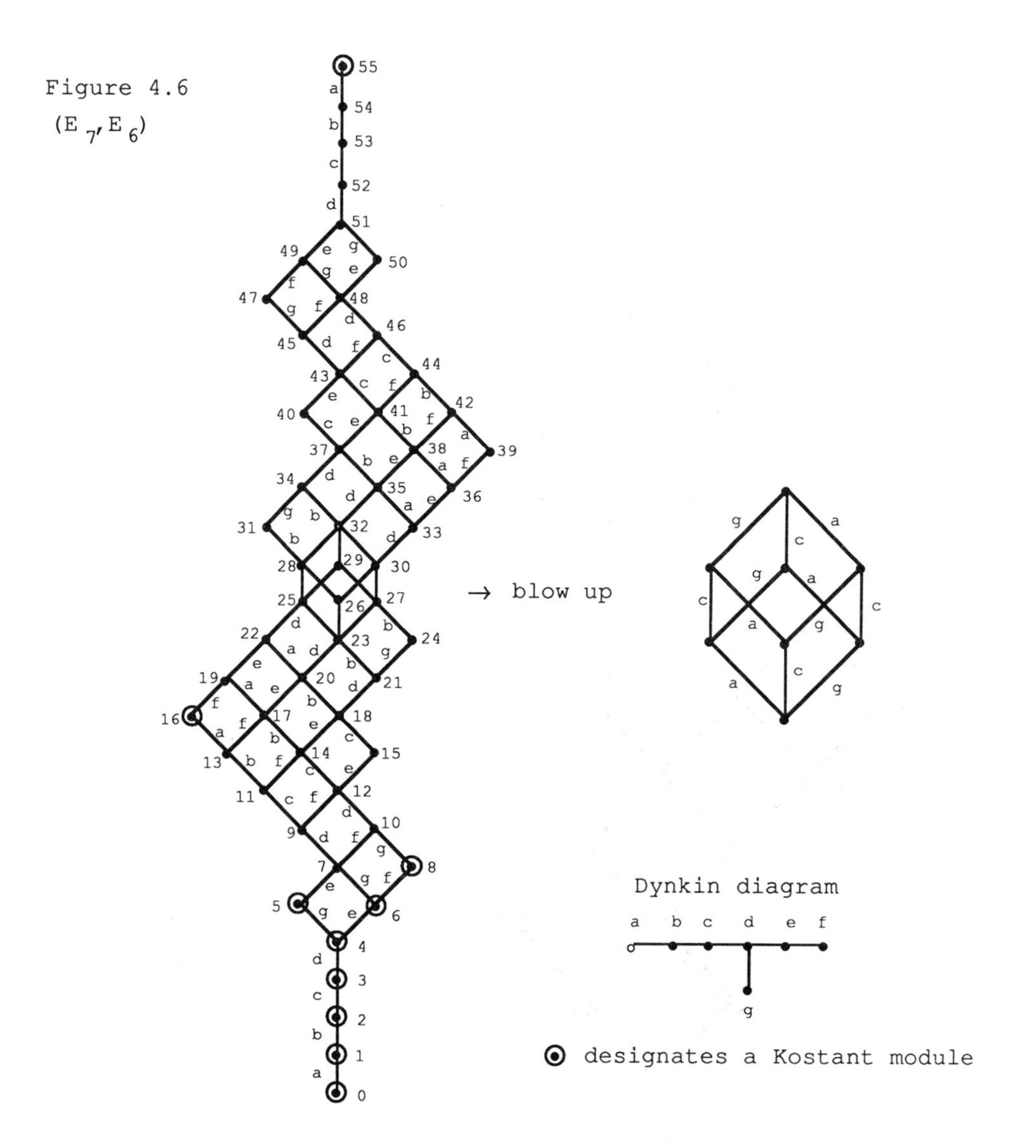

Figure 4.6
(E_7, E_6)

§5. Notational Conventions and a Lemma

In this section we set down some of the algebraic consequences for the class of modules which will be the focus of the remainder of the article. Let F be a sum of irreducible highest weight $\mathfrak{g}$-modules F_i, $i \in I$, and let H_i be the $\mathfrak{k}$-submodule of F_i generated by the highest weight vector. Put H equal to the sum of the H_i. In our main applications F will be Fock space and H will be the subspace of harmonic polynomials. However, for the moment, this additional structure is unnecessary.

We keep in force convention (1.5) by which bold face type is used to designate vector valued polynomials on $\mathfrak{p}_+$; e.g. $\mathbf{H} = P(\mathfrak{p}_+, H)$. Since $\mathfrak{p}_+$ acts by zero on H, $\mathbf{H}$ has a $\mathfrak{g}$-module action $\mathfrak{t}^*$ when canonically identified with $U(\mathfrak{g}) \otimes_{U(\mathfrak{k}\oplus\mathfrak{p}_+)} H$. Define the radical of a module to be the intersection of its maximal submodules. Put $J(H)$ equal to the radical of $(\mathfrak{t}^*, \mathbf{H})$.

Since H is a direct sum of finite dimensional irreducible $\mathfrak{k}$-modules, $\mathbf{H}$ is the direct sum of generalized Verma modules and $J(H)$ is the direct sum of the maximal submodules of these generalized Verma modules. For later reference we record as a lemma the following consequence of this observation.

5.1 LEMMA. Let V be a $\mathfrak{k}$-submodule of H and let P_V be a $\mathfrak{k}$-module projection of H onto V. Let $\mathbf{P}_V$ denote the induced map $\mathbf{P}_V{:}\mathbf{H} \to \mathbf{V}$ defined by $F(T) \to P_V(F(T))$. Then $\mathbf{P}_V$ is a $\mathfrak{g}$-module projection of $\mathbf{H}$ onto $\mathbf{V}$ carrying $J(H)$ onto $J(V)$.

Let $\mathbb{F}$ and J(F) be defined as above with F replacing H. Since $(\mathfrak{t}^*, \mathbb{F})$ is an induced module, the identity map of F extends uniquely to a $\mathfrak{g}$-module map Θ: $\mathbb{F} \to F$ given by $x \otimes f \to x \cdot f$, $x \in U(\mathfrak{g})$, $f \in F$. Again, for later reference we record the following lemma.

5.2 LEMMA. $J(H) = \mathbb{H} \cap \text{kernel } \Theta$

PROOF. Since H is the direct sum of the $\mathfrak{k}$-submodules H_i, $\mathbb{H}$ is the direct sum of the $\mathbb{H}_i$. Moreover, the inclusion of H_i into F_i is a $\mathfrak{k} \oplus \mathfrak{p}_+$-module map and so extends uniquely to a $\mathfrak{g}$-module map Θ_i of $\mathbb{H}_i$ to F_i. By irreducibility of F_i, Θ_i is onto and kernel Θ_i = $J(H_i)$. Clearly, Θ restricted to $\mathbb{H}$ is the sum of the Θ_i which proves the lemma.

§6. The Cone Decomposition

Some ten years ago the unitary highest weight modules were classified in [15] and [27]. In this section we briefly recall the parameterization given by the first of these articles and indicate how it leads to what we call the cone decomposition. The cone decomposition and in particular the vertices of the cones will play an important role in this section as well as in much of what follows.

Let $\mathfrak{h}_{\mathbb{R}}^*$ denote the real span of the roots. For us a cone with vertex zero (in $\mathfrak{h}_{\mathbb{R}}^*$) will be the intersection of a (nonempty) collection of closed half spaces. So each cone C is determined by a finite set $\{h_i\}$ of elements of $\mathfrak{h}$ with $C = \{\upsilon \in \mathfrak{h}_{\mathbb{R}}^* : \upsilon(h_i) \geq 0$ for all i$\}$. An integral cone will be the intersection of a cone with the set of $\mathfrak{k}$-integral points of $\mathfrak{h}^*$. For υ an integral element in $\mathfrak{h}^*$, a translated integral cone (TIC) with vertex υ is a set of the form $\upsilon + C$ with C an integral cone.

We now summarize the classification given in [15]. Let Λ denote the subset of $\mathfrak{h}^*$ of highest weights of unitarizable highest weight modules and let Λ_r denote those points λ in Λ for which $N(\lambda + \rho)$ is reducible. From the point of view of this article, Λ_r is the more interesting set since this is the set from which the differential equations arising in (2.9) are not vacuous. We shall call it the ***set of reduction points***.

Assume G is simple. Let β denote the unique maximal root in Δ^+ and define $\zeta \in \mathfrak{h}^*$ with ζ orthogonal to the compact roots and

$\zeta_\beta = 2(\zeta,\beta)/(\beta,\beta) = 1$. Consider the lines $L(\lambda) = \{\lambda + z\zeta : z \in \mathbb{R}\}$, $\lambda \in \mathfrak{h}^*$ and $\mathfrak{k}$-dominant integral. A normalization is chosen for each line so that $z = 0$ corresponds to the unique point with highest weight module a limit of discrete series module. When λ is such we write λ_0 in place of λ and the line in the form $\{\lambda_0 + z\zeta : z \in \mathbb{R}\}$. Then $(\lambda_0 + \rho, \beta) = 0$ and the set of values z with $\lambda = \lambda_0 + z\zeta \in \Lambda$ is of the form:

(6.1)

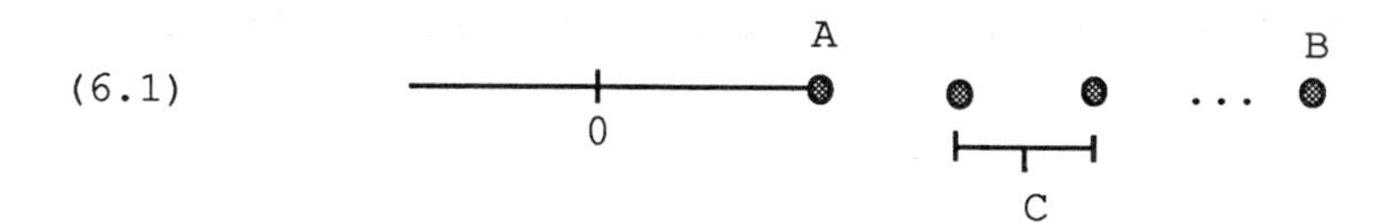

The points corresponding to elements of Λ_r are precisely the equally spaced (at intervals of length C) points from A to B. The constant C depends only on $\mathfrak{g}$ while A and B depend on the line $L(\lambda)$ as well. When it is important to emphasize this dependence we write $A = A(\lambda)$ and $B = B(\lambda)$. The values of C are given in Table (6.20) below.

Associated to each line $L(\lambda)$ we have two real root systems $Q(\lambda)$ and $R(\lambda)$ defined as in [15] with $Q(\lambda)$ contained in $R(\lambda)$. If $\lambda \in \Lambda_r$ the natural grading of polynomials in $\mathfrak{p}_+$ gives a grading of the maximal submodule J_λ. Let $\ell(\lambda)$ denote the degree of polynomials of least degree in J_λ. We refer to the integer $\ell(\lambda)$ as the ***level of reduction*** of $N(\lambda + \rho)$ [cf. (3.10ii)]. When $\lambda = \lambda_0 + z\zeta$, we shall frequently write $\ell(z)$ in place of $\ell(\lambda)$. We caution the reader that in [15] the symbol $N(\lambda)$ is written in place of $N(\lambda+\rho)$.

6.2 DEFINITION. For $\lambda \in \Lambda_r$, let $a(\lambda) = (Q(\lambda), R(\lambda), \ell(\lambda))$ and let $\mathcal{A}$ denote the set of all such triples as λ ranges over Λ_r. For $a \in \mathcal{A}$, let Λ_a denote the set of all $\lambda \in \Lambda_r$ with $a(\lambda) = a$.

The notion of level of reduction was fundamental to the classification in [15] and the connection with the parameters in (6.1) is given by the following lemma.

6.3 LEMMA. With notation as above we write $\lambda = \lambda_0 + z\zeta$ and put $r(\lambda)$ equal to the split rank of root system $Q(\lambda)$. Then $B - A = C(r(\lambda) - 1)$ and for non-negative integers $n < r(\lambda)$, $\ell(B - nC) = n + 1$.

PROOF. This result is implicit in [15]. The number of reduction points on the line $L(\lambda)$ equals the split rank of $Q(\lambda)$. Furthermore the level of reduction $\ell(A)$ equals this split rank and so Proposition 3.15 [15] implies that the level of reduction drops by one at each point from A to B, giving the formula in the lemma.

6.4 COROLLARY. Let $a = (Q,R,\ell) \in \mathcal{A}$ and let $\lambda \in \Lambda_a$. If we write $\lambda = \lambda_0 + z\zeta$ as above, then $z = B(\lambda) - (\ell-1)C$.

6.5 DEFINITION. For $a = (Q,R,\ell) \in \mathcal{A}$, let C_a be the integral cone of $\mathfrak{k}$-dominant integral elements in $\mathfrak{h}^*_{\mathbb{R}}$ which are orthogonal to elements in R. The decomposition of Λ_r is now given as:

6.6 PROPOSITION.

i) Λ_r is the disjoint union of the sets Λ_a, $a \in \mathcal{A}$

ii) Let λ_a be defined by Table 6.22 below. Then Λ_a is a translated integral cone with vertex λ_a and $\Lambda_a = \lambda_a + C_a$.

PROOF. Part i) of the proposition follows directly from our definitions while ii) is established most easily by a case by case verification following the notation used in sections seven through thirteen in [15]. We omit the detailed verification.

Next we turn to a description and formula for the highest weight of the maximal submodule J_λ. We begin by recalling some of the notation and results from [15].

Following Harish-Chandra define a collection of orthogonal roots as follows for the positive root system $-\Delta_n^+ \cup \Delta_c^+$. Let ξ_1 be the least element in $-\Delta_n^+$ (i.e. $\xi_1 = -\beta$ where β is the maximal root in Δ_n^+). Inductively define ξ_k to be the least element in $-\Delta_n^+$ which is

orthogonal to $\xi_1,\ldots,\xi_{k-1}$. Let $\xi_1,\ldots,\xi_r$ be the maximal collection obtained. Then r is the split rank of $\mathfrak{g}_0$.

As is well known these roots describe the $\mathfrak{k}$-module decomposition of $S(\mathfrak{p}_-)$ as follows:

6.7 THEOREM. (Schmid [43]) Let I denote the set of integral multi-indices $\underline{i} = (i_1,\ldots,i_r)$ with $i_1 \geq i_2 \geq \ldots \geq i_r \geq 0$. Let $F_{\underline{i}}$ denote the irreducible finite dimensional $\mathfrak{k}$-module with lowest weight $\underline{i}\cdot\xi = \sum_{1\leq j\leq r} i_j\xi_j$. Then $S(\mathfrak{p}_-)$ has a multiplicity free decomposition

$$S(\mathfrak{p}_-) \simeq \bigoplus_{\underline{i}\in I} F_{\underline{i}} .$$

For convenience we put $\nu_i = \xi_1 + \ldots + \xi_i$, for $1 \leq i \leq r$.

For two irreducible finite dimensional $\mathfrak{k}$-modules define the PRV component (of the tensor product) to be the unique irreducible submodule with extreme weight equal to the highest weight of one factor plus the lowest weight of the other factor. The existence of the PRV component is established in [40].

6.8 PROPOSITION. [13] Suppose $\lambda \in \Lambda_a$ with $a = (Q, R, \ell)$. Let u and v denote respectively the unique elements of maximal length in the Weyl groups for the positive root systems $Q \cap \Delta_c^+$ and $R \cap \Delta_c^+$ and let μ denote the highest weight of the maximal submodule J_λ. Then

$$\mu = \lambda + \frac{1}{2}(u\nu_\ell + v\nu_\ell)$$

in all cases except when $\mathfrak{g}_0 \simeq SO(2,2n-1)$ and $Q \neq R$. In this case

$$\mu = \lambda + \frac{1}{2}(\nu_\ell + v\nu_\ell) \ .$$

Moreover, in all cases $F(\mu)$ occurs with multiplicity one in $N(\lambda+\rho)$ and in all cases $F(\mu)$ is a PRV component of a tensor product in $S(\mathfrak{p}_-) \otimes F(\lambda)$.

This result is established in [13] as part of an analysis of unitarity itself. For the convenience of the reader we include an alternate proof along the lines of the presentation here.

PROOF. Grade $N = N(\lambda+\rho) \simeq S(\mathfrak{p}_-) \otimes F(\lambda)$ by the grading of homogeneous tensors in $S(\mathfrak{p}_-)$. By assumption the level of reduction is ℓ and so by Proposition 3.9 [15], for all $\mathfrak{k}$-modules $F(\xi)$ occurring in N in degree less than ℓ, $\|\xi+\rho\| > \|\lambda+\rho\|$. Moreover by Lemma 3.6 [15], there exists exactly one ξ with $F(\xi)$ occurring in N in degree ℓ with $\|\xi+\rho\| = \|\lambda+\rho\|$. Moreover, for this ξ, $\xi = \mu$. Let $\mu_1 = \lambda + \frac{1}{2}(u\nu_\ell + v\nu_\ell)$. To prove (6.8) it is sufficient to establish two points:

(6.9) $F(\mu_1)$ occurs in N in degree ℓ

and

(6.10) $$\|\mu_1 + \rho\| = \|\lambda + \rho\| .$$

First suppose that $Q = R$. Then from the table in [13] we see that $\lambda + u\nu_\ell$ is Δ_c^+-dominant. So $\mu_1 = \lambda + u\nu_\ell = u(\lambda+\nu_\ell)$. Since λ is a highest weight and ν_ℓ is a lowest weight, the PRV theorem [40] applied to (6.7) shows that $F(\mu_1)$ occurs in N in degree ℓ. We now verify identity (6.10).

$$\|\mu_1+\rho\|^2 - \|\lambda+\rho\|^2 = 2(\lambda+\rho, u\nu_\ell) + (\nu_\ell, \nu_\ell)$$

(6.11) $$= 2(\lambda+u\rho, \nu_\ell) + \ell(\xi_1, \xi_1)$$

$$= 2(\lambda+\rho, \nu_\ell) - 4(\rho_{Q,c}, \nu_\ell) + \ell(\xi_1, \xi_1)$$

where $\rho_{Q,c}$ is half the sum of the roots in $Q \cap \Delta_c^+$.

Recall that $\nu_\ell = \xi_1 + \ldots + \xi_\ell$ and put $\theta_i = \xi_i - \xi_1$. Then θ_i is a sum of elements from $\Delta_c^+ \cap Q$. For any $\zeta \in \mathfrak{g}^*$ define the *normalized height* of ζ as:

$$\underset{\sim}{h}(\zeta) = 2(\rho, \zeta)/(\xi_1, \xi_1) .$$

6.12 LEMMA. For $1 \le i \le \ell$,

$$2(\rho_{Q,c}, \theta_i)/(\xi_1, \xi_1) = \underset{\sim}{h}(\theta_i) = 2(i-1)C$$

where C is given in Table (6.20).

PROOF. The first identity follows since θ_i is a sum of simple roots in $Q \cap \Delta_c^+$. The second identity can be easily verified case by case. We include only the calculation for E_7 to indicate the method. Let the Dynkin diagram be labelled as usual

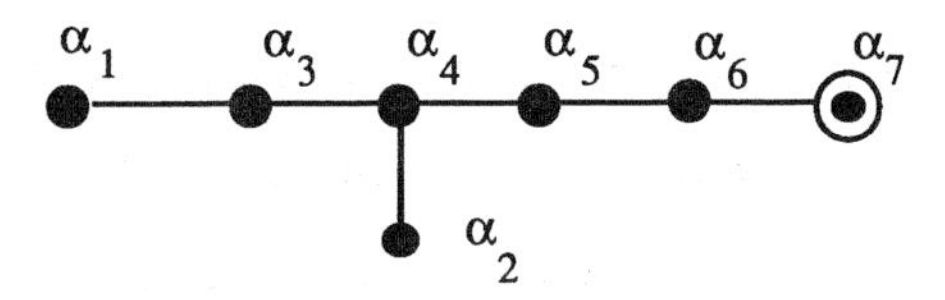

Then for the calculation of ξ_i, $1 \leq i \leq 3$, let the α_i, $1 \leq i \leq 6$, be compact and $\alpha_7 = \xi_1$. One verifies that $\xi_2 = \alpha_7 + 2\alpha_6 + 2\alpha_5 + 2\alpha_4 + \alpha_3 + \alpha_2$ and ξ_3 equals the maximal root in this system $\alpha_7 + 2\alpha_6 + 3\alpha_5 + 4\alpha_4 + 3\alpha_3 + 2\alpha_2 + 2\alpha_1$. In turn $\underset{\sim}{h}(\theta_i) = 8$ and 16 for $i = 2$ and 3. This proves the lemma in this case.

We now return to the calculation in (6.11). From (6.4) and Theorem 2.8(b) [15] we have:

$$(6.13) \qquad (\lambda+\rho,\beta^\vee) = B(\lambda) - (\ell-1)C = 1 + 2(\rho_{Q,c},\beta^\vee) - (\ell-1)C .$$

Thus

$$\begin{aligned}
\|\mu_1+\rho\|^2 - \|\lambda+\rho\|^2 &= \sum_{1\leq i\leq \ell} 2(\lambda+\rho,\xi_i) - 4(\rho_{Q,c},\xi_i) + \ell(\xi_1,\xi_1) \\
&= 2\ell(\lambda+\rho,\xi_1) - 4\ell(\rho_{Q,c},\xi_1) + \ell(\xi_1,\xi_1) \\
&\quad + \sum_{1\leq i\leq \ell} 2(\rho,\theta_i) - 4(\rho_{Q,c},\theta_i) \\
(6.14) \qquad &= \ell(\ell-1)C\ (\xi_1,\xi_1) && \text{by (6.13)} \\
&\quad + \sum_{1\leq i\leq \ell} -2(i-1)C\ (\xi_1,\xi_1) && \text{by (6.12)} \\
&= 0 .
\end{aligned}$$

This completes the proof of (6.8) for the cases with $Q = R$.

Now suppose $Q \neq R$. Here there are but two cases and the results differ slightly. So we treat the cases separately. First suppose $Q \neq R$ and $\mathfrak{g}_0 \simeq sp(2n,\mathbb{R})$. We shall write elements of $\mathfrak{h}^*$ in the usual Euclidean coordinates as in [15]. Choose q and r with $a = (Q,R,\ell)$ and Q and R root systems of $sp(2q,\mathbb{R})$ and $sp(2r,\mathbb{R})$. Set $t = \frac{1}{2}(q+r)$ and let $\zeta = (1,\ldots,1)$. Then $\lambda + \rho$ has the form $(t,t-1,\ldots,\widehat{\left(\frac{r-q}{2}\right)},\ldots,\left(\frac{r-q}{2}\right), \underline{a}) - \left(\frac{\ell-1}{2}\right)\zeta$, where the symbol $\widehat{\ }$ denotes omission of this value as a coordinate, $\underline{a}$ is an (n-r)-tuple of strictly decreasing integers (or half integers) and $a_1 \leq \frac{q-r}{2} - 2$.

Now ν_ℓ equals the n-tuple whose first ℓ-coordinates are -2 with the remaining ones zero. The elements u and v are permutations with $u\nu_\ell$ (resp. $v\nu_\ell$) having -2 in the $(q+1-\ell)^{th}$ through q^{th} (resp. $(r+1-\ell)^{th}$ through r^{th}) coordinates. A short calculation (which we omit) shows that μ_1 lies in the $\mathcal{W}_c$-orbit of $\lambda+\nu_\ell$ and so $F(\mu_1)$ is the PRV-component in $F_{\underline{i}} \otimes F(\lambda)$ where $\underline{i}$ is the r-tuple whose first ℓ-coordinates are ones with the remaining coordinates zero and $F_{\underline{i}}$ is defined as in (6.7). This proves (6.9) for this case.

We now turn to the norm calculation for $\mu_1 + \rho$. Consider the subset of the n-tuple $\lambda + \rho$ including the entries of the $(q+1-\ell)^{th}$ through r^{th} coordinates. Put $d = r - q + \ell$. Then this d-tuple $\nu = (\nu_1,\ldots,\nu_d)$ has the form $(a+1,a,\ldots,\hat{j},\ldots,-a)$ where a is a positive integer or half integer, $\hat{j}$ denotes omission of the value j and $j \neq a+1, \neq a$. Set $\xi = \frac{1}{2}(u\nu_\ell + v\nu_\ell)$ then the nonzero coordinates of ξ lie within the same range, $(q+1-\ell)^{th}$ through r^{th} coordinates. Let ξ' denote this subset d-tuple of ξ. The omitted value j equals $a+1-\ell$ and thus $\nu+\xi'$ equals $(a,\ldots,\widehat{-j},\ldots,-a-1)$. From this we conclude that the n-tuple $\mu_1+\rho$ has as its set of coordinates exactly those of $\lambda+\rho$ with the exceptions that the coordinates $a+1$ and $-j$ of $\lambda+\rho$ are replaced by $-a-1$ and j in $\mu_1+\rho$. So they are equal up to sign and thus have the same norm. This proves (6.10) and completes the proof of (6.8) for this case.

The remaining case is $\mathfrak{g}_0 \simeq so(2,2n-1)$ and $Q \neq R$. Here we have only one point $\lambda+\rho = (1/2,n-1,n-2,\ldots,1)$. The level of reduction is one and so Lemma 4.11 [15] implies that $\mu+\rho = (-1/2,n-1,n-2,\ldots,1)$. This gives the formula in (6.8) for this last case.

The rest of this section is devoted to the statement and proof of the Factorization Theorem. This will offer the first application of the cone decomposition to the representation theory at hand.

Fix $a \in \mathcal{A}$ and let $\lambda = \lambda_a + \lambda' \in \Lambda_a$. Let H_a, H' and H_λ denote $\mathfrak{k}$-irreducible spaces with highest weights λ_a, λ' and λ, respectively. Following the notation of (3.1) we let J_a and J_λ denote the maximal submodules of $\mathbb{H}_a$ and $\mathbb{H}_\lambda$ respectively. Then $\overset{o}{J}_a$ (resp. $\overset{o}{J}_\lambda$) denotes the irreducible $\mathfrak{k}$-module generating J_a (resp. J_λ) as a $U(\mathfrak{p}_-)$-module. Put $E = H_a \otimes H'$ and let $P:E \to H_\lambda$ be a nonzero $\mathfrak{k}$-equivariant map. Thus P is surjective. We define an extension $\mathbb{P}:\mathbb{H}_a \otimes H' \to \mathbb{H}_\lambda$ by the formula $\mathbb{P}(F \otimes v)(T) = P(F(T) \otimes v)$. The space $\mathbb{E} = P(\mathfrak{p}_+,E)$ carries a canonical $\mathfrak{g}$-structure and moreover the map $\mathbb{P}$ above is precisely (cf. §5) the $\mathfrak{g}$-module map induced from the $\mathfrak{k}$-module map $P:E \to H_\lambda$. Consider $J_a \otimes H'$ as a subspace of the $\mathfrak{g}$-module $\mathbb{E}$. Although not $\mathfrak{g}$-stable, this space is a $\bar{\mathfrak{q}}$-module and so the restriction of P to it is a $\bar{\mathfrak{q}}$-module homomorphism.

6.15 FACTORIZATION THEOREM. With notation as above let $\lambda = \lambda_a + \lambda'$. Set μ_a (resp. μ_λ) equal to the highest weight of J_a^o (resp. J_λ^o). Then

(1) $\mu_\lambda = \mu_a + \lambda'$

(2) $P(J_a^o \otimes H') = J_\lambda^o$ and

(3) $P(J_a \otimes H') = J_\lambda$.

6.16 REMARKS 1. In this generality the factorization given in (2) and (3) is realized by the map induced from the $\mathfrak{k}$-homomorphism P. In later sections we will specialize to cases where this map is given explicitly as multiplication of functions and identity (2) is a consequence of the theory of Young products. In fact the term factorization is suggested by this realization (cf. 8.5).

2. The factorization theorem was proved in the harmonic setting by Davidson and Stanke. The general statement above, its proof, and the formula for the remaining μ_a as given in Table (6.22) were first provided by Davidson. The observation by Stanke that (6.15.1) implies (6.15.2) and Proposition (6.8) lead to the present form of the proof.

3. Let $\mathcal{B}_\ell$ denote the set of all sums of ℓ positive noncompact roots. If $a = (Q, R, \ell)$ and we write $\mu_\lambda = \lambda - B_\lambda$ and $\mu_a = \lambda_a - B_a$ where $B_a, B_\lambda \in \mathcal{B}_\ell$, then (1) of (6.15) implies $B_\lambda = B_a$ for all $\lambda \in \Lambda_a$.

PROOF. Identity (1) follows directly from (6.8). We now turn to the proof of (2) which will rely on the following lemmas.

6.17 LEMMA. The restriction of P to $J_a^o \otimes H'$ is nonzero.

PROOF. Let M be the span of $\{F(T): T \in \mathfrak{p}_+, F \in J_a^o\}$. Since $M \neq 0$ and is $\mathfrak{k}$-invariant, we have $M = H_a$. Now since $\mathrm{P}(F \otimes v)(T) = P(F(T) \otimes v)$, we have $\{\mathrm{P}(F \otimes v)(T): T \in \mathfrak{p}_+, F \in J_a^o, v \in H'\} = P(E) = H_\lambda$.

6.18 LEMMA. Let $B, B' \in \mathcal{B}_\ell$ (cf. (6.16)(3)) such that

1) $2(\lambda+\rho,B') = (B',B')$ and

2) $\lambda - B$ is $\mathfrak{k}$-dominant.

If $B = B' + \nu$ with ν the (possibly empty) sum of positive compact roots, then $2(\lambda+\rho,B) \geq (B,B)$ where equality holds if and only if $B = B'$.

PROOF. We have $2(\lambda+\rho,B) = 2(\lambda+\rho,B') + 2(\lambda+\rho,\nu) = (B',B') + 2(\lambda+\rho,\nu)$. Now observe that $2(\lambda-B+\rho,\nu) = 2(\lambda-B,\nu) + 2(\rho,\nu)$ and $(\lambda-B,\nu) \geq 0$ by 2) and $(\rho,\nu) > 0$ unless $\nu = 0$. Thus

$$2(\lambda+\rho,v) \geq 2(B,\nu) = 2(B',\nu) + 2(\nu,\nu) \geq 2(B',\nu) + (\nu,\nu).$$

We then have

$$2(\lambda+\rho,B) \geq (B',B') + 2(B',\nu) + (\nu,\nu) = (B,B)$$

with equality if and only if $\nu = 0$.

We let $\mathbb{H}_\lambda^\ell$ denote the space of homogeneous polynomials in $\mathbb{H}_\lambda = P(\mathfrak{p}_+,H_\lambda)$ of degree ℓ.

6.19 LEMMA. Let $\lambda = \lambda_a + \lambda' \in \Lambda_a$ with $a = (Q,R,\ell)$. Set $A = \mathrm{Hom}_{\mathfrak{k}}(\overset{\circ}{J}_a \otimes H', H_\lambda^\ell)$. Then A has dimension one and is spanned by the restriction of P to $\overset{\circ}{J}_a \otimes H'$. Moreover, $\mathsf{P}(\overset{\circ}{J}_a \otimes H') = \overset{\circ}{J}_\lambda$.

PROOF. Suppose Φ is a nonzero element of A and let v be a highest weight vector in $\overset{\circ}{J}_a \otimes H'$ with highest weight ξ and $\Phi(v) \neq 0$. Then $\xi = \mu_a + \omega = \lambda_a - B_a + \omega$ where ω is a weight of H'. On the other hand, all highest weights of H_λ^ℓ are of the form $\lambda - B$, $B \in \mathcal{B}_\ell$. Since ℓ is the level of reduction for $N(\lambda+\rho)$ either $\Phi(v)$ is a Δ^+-highest weight of the maximal submodule or ξ is a $\mathfrak{k}$-highest weight of $L(\lambda+\rho)$. Then by Lemma 3.6 and Proposition 3.9 in [15] we have $|\xi+\rho| \geq |\lambda+\rho|$ which gives $2(\lambda+\rho, B) \leq (B,B)$. Note that $B = B_a + \lambda' - \omega$. Consider (6.18) and set $B' = B_a$ and $\nu = \lambda' - \omega$. By (1) of (6.15) and (6.8) $|\lambda - B_a + \rho| = |\lambda+\rho|$. This implies the validity of hypothesis 1) in (6.18). Moreover the condition on ν holds and so $B = B_a$ and $\xi = \lambda - B_a$. By Lemma 3.6 in [15], $\Phi(v)$ is a Δ^+-highest weight vector in $N(\lambda+\rho)$ and so $\Phi(v)$ lies in the maximal submodule. But consideration of the level of reduction implies $\Phi(v) \in \overset{\circ}{J}_\lambda$ so that the image of Φ is $\overset{\circ}{J}_\lambda$. The identity (1) in (6.15) implies that $\lambda - B_a$ is the sum of the highest weights of $\overset{\circ}{J}_a$ and H'. Since this $\mathfrak{k}$-type occurs in $\overset{\circ}{J}_a \otimes H'$ with multiplicity one, we have $\dim(A) \leq 1$. By (6.17) we know P restricts to a nonzero element of A, which completes the proof.

Lemma (6.19) then completes the proof of (6.15)(2). Identity (3) follows from (2) since P is a $\bar{\mathfrak{q}}$-module homomorphism.

Table 6.20

$\mathfrak{g}_0$	SU(p,q)	Sp(n,ℝ)	SO*(2n)	SO(2,2n-2)	SO(2,2n-1)	EIII	EVII
C	1	1/2	2	n-2	n-3/2	3	4

In Table 6.21 below we give the cones C_a for $a \in \mathcal{A}$ and G = SU(p,q), Sp(n,ℝ) and SO*(2m). The coefficients a_i denote nonnegative integers and the weight ω_i designates the i-th fundamental weight associated to the standard Dynkin diagrams as labeled in sections seven through thirteen of [15]. For example, if G = Sp(n,ℝ), then $\omega_i = \sum_{j=1}^{i} e_j$.

In Table 6.22, the weights λ_a and μ_a are listed for all $a \in \mathcal{A}$. All highest weights μ_λ may be then found using (6.15)(1). Note that the root system R is given only when Q ≠ R. In the formulas for μ_a if an index j occurs for which ω_j is undefined then by convention that term is omitted from the expression. Finally, the case m = 3 for SO*(2m) is intentionally excluded from Table 6.22. The relevant data for SO*(6) can be obtained from the line in the table for SU(1,3) since these groups are locally isomorphic.

TABLE 6.21

SU(p,q): Let $a = (Q,R,\ell), Q = R = SU(p',q'), 1 \leq p' \leq p$ and $1 \leq q' \leq q$. Then

$$C_a = \{a_{p'}\omega_{p'} + \ldots + a_p\omega_p + \ldots + a_{n-q'}\omega_{n-q'}:$$

$$a_p = - a_{p'} - \ldots - a_{p-1} - a_{p+1} - \ldots - a_{n-q'}\}.$$

Sp(n,ℝ): Let $a = (Q,R,\ell)$, $R = Sp(r,\mathbb{R})$, and $1 \leq \ell \leq r \leq n$. Then

$$C_a = (a_r\omega_r + \ldots + a_n\omega_n: \ a_n = -(a_r + \ldots + a_{n-1})).$$

SO*(2m): Let $a = (Q,R,\ell)$. $Q = R$

Case 1: $R = SO^*(2p)$, $3 \leq p \leq m$

$$C_a = \{a_p\omega_p + \ldots + a_{m-1}\omega_{m-1} + a_m\omega_m:$$

$$a_m = -2a_p - \ldots - 2a_{m-2} - a_{m-1}\}$$

Case 2: $R = SU(1,q)$, $1 \leq q \leq m - 1$

$$C_a = \{a_1\omega_1 + a_{q+1}\omega_{q+1} + \ldots + a_{m-1}\omega_{m-1} + a_m\omega_m:$$

$$a_m = -(a_1 + 2a_{q+1} + \ldots + 2a_{m-2} + a_{m-1})\}.$$

TABLE 6.22

Group	Vertex λ_a	Weight μ_a	Parameters	Q	R	$\ell(\lambda_a)$
SU(p,q)	$\omega_{p'} + \omega_{n-q'}$	$\omega_{p'-\ell} + \omega_{n-q'+\ell}$	$1 \leq p' \leq p$,	SU(p',q')	–	$1 \leq \ell \leq$
$p \geq 1, q \geq 1$	$-(n+\ell+1-p'-q')\omega_p$	$-(n+\ell+1-p'-q')\omega_p$	$1 \leq q' \leq q$			$\min(p',q')$
Sp(n,ℝ)	$\omega_q + \omega_r$	$\omega_{q-\ell} + \omega_{r-\ell}$	$1 \leq q \leq r \leq n$	Sp(q,ℝ)	Sp(r,ℝ)	$1 \leq \ell \leq q$
$n \geq 1$	$-(2+n-\frac{1}{2}(r+q-\ell+1))\omega_n$	$-(2+n-\frac{1}{2}(r+q-\ell+1))\omega_n$				
SO*(2m)	$\omega_2 - (2m-2)\omega_m$	$-(2m-2)\omega_m$		SU(1,1)	–	1
$m \geq 4$	$\omega_p - 2(m-p+\ell)\omega_m$	$\omega_{p-2\ell} - 2(m-p+\ell)\omega_m$	$3 \leq p \leq m-2$	SO*(2p)	–	$1 \leq \ell \leq [p/2]$
	$\omega_{m-1} - (1+2\ell)\omega_m$	$\omega_{m-1-2\ell} - 2(1+\ell)\omega_m$		SO*(2m-2)	–	$1 \leq \ell \leq [\frac{m-1}{2}]$
	$-(2\ell-2)\omega_m$	$\omega_{m-2\ell} - 2\ell\omega_m$		SO*(2m)	–	$1 \leq \ell \leq [m/2]$
	$\omega_1 + \omega_{q+1} - (2m-q)\omega_m$	$\omega_q - (2m-q)\omega_m$	$2 \leq q \leq m-3$	SU(1,q)	–	1
	$\omega_1 + \omega_{m-1} - (m+1)\omega_m$	$\omega_{m-2} - (m+2)\omega_m$		SU(1,m-2)	–	1
	$\omega_1 - (m-1)\omega_m$	$\omega_{m-1} - m\omega_m$		SU(1,m-1)	–	1

TABLE 6.22

G	Vertex λ_a	Weight μ_a	Parameters	Q	R	$\ell(\lambda_a)$
SO(2,2m-2)	$-(2m-p-1)\omega_1+\omega_{p+1}$	$-(2m-p)\omega_1+\omega_p$	$1 \le p \le m-3$	SU(1,p)	-	1
m≥3	$-(m+1)\omega_1+\omega_{m-1}+\omega_m$	$-(m+2)\omega_1+\omega_{m-2}$		SU(1,m-2)	-	1
	$-(m-1)\omega_1+\omega_m$ (1)	$-m\omega_1+\omega_{m-1}$		SU(1,m-1)	-	1
	$-(m-2)\omega_1$	$-m\omega_1$		SO(2,2m-2)	-	2
	0	$-2\omega_1+\omega_2$		SO(2,2m-2)	-	1
	$-(m-1)\omega_1+\omega_{m-1}$ (2)	$-m\omega_1+\omega_m$		SU(1,m-1)	-	1
SO(2,2m-1)	$-(2m-p)\omega_1+\omega_{p+1}$	$-(2m-p+1)\omega_1+\omega_p$	$1 \le p \le m-2$	SU(1,p)	-	1
m≥2	$-(m+1)\omega_1+2\omega_m$	$-(m+2)\omega_1+\omega_{m-1}$		SU(1,m-1)	-	1
	0	$-2\omega_1+\omega_2$		SO(2,2m-1)	-	1
	$-(m-3/2)\omega_1$	$-(m+1/2)\omega_1$		SO(2,2m-1)	-	2
	$-(m-1/2)\omega_1+\omega_m$	$-(m+1/2)\omega_1+\omega_m$		SU(1,m-1)	SO(2,2m-1)	1

(1) Dynkin diagram

(2) Dynkin diagram

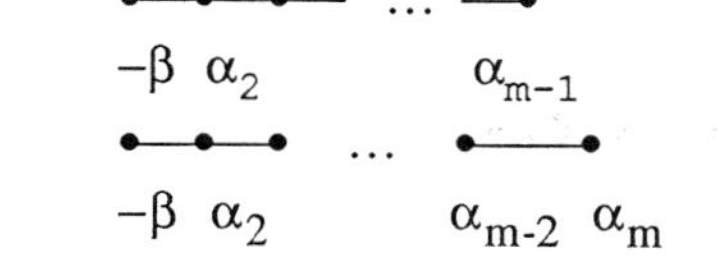

TABLE 6.22

G	Vertex λ_a	Weight μ_a	$Q = R$	$\ell(\lambda_a)$
EIII	$-12\omega_1 + \omega_2$	$-12\omega_1$	SU(1,1)	1
	$-12\omega_1 + \omega_4$	$-12\omega_1 + \omega_2$	SU(1,2)	1
	$-12\omega_1 + \omega_3 + \omega_5$	$-12\omega_1 + \omega_4$	SU(1,3)	1
	$-9\omega_1 + \omega_5$ (1)	$-10\omega_1 + \omega_3$	SU(1,4)	1
	$-10\omega_1 + \omega_3 + \omega_6$ (2)	$-10\omega_1 + \omega_5$	SU(1,4)	1
	$-8\omega_1 + \omega_3$	$-8\omega_1 + \omega_6$	SU(1,5)	1
	$-5\omega_1 + \omega_6$	$-6\omega_1 + \omega_2$	SO(2,8)	1
	$-8\omega_1 + \omega_6$	$-9\omega_1$	SO(2,8)	2
	0	$-2\omega_1 + \omega_3$	EIII	1
	$-3\omega_1$	$-5\omega_1 + \omega_6$	EIII	2
EVII	$\omega_1 - 18\omega_7$	$-18\omega_7$	SU(1,1)	1
	$\omega_3 - 18\omega_7$	$\omega_1 - 18\omega_7$	SU(1,2)	1
	$\omega_4 - 18\omega_7$	$\omega_3 - 18\omega_7$	SU(1,3)	1
	$\omega_2 + \omega_5 - 18\omega_7$	$\omega_4 - 18\omega_7$	SU(1,4)	1
	$\omega_5 - 15\omega_7$ (3)	$\omega_2 - 15\omega_7$	SU(1,5)	1
	$\omega_2 + \omega_6 - 16\omega_7$ (4)	$\omega_5 - 16\omega_7$	SU(1,5)	1
	$\omega_2 - 13\omega_7$	$\omega_6 - 14\omega_7$	SU(1,6)	1
	$\omega_6 - 10\omega_7$	$\omega_1 - 10\omega_7$	SO(2,10)	1
	$\omega_6 - 14\omega_7$	$-14\omega_7$	SO(2,10)	2
	0	$\omega_6 - 2\omega_7$	EVII	1
	$-4\omega_7$	$\omega_1 - 6\omega_7$	EVII	2
	$-8\omega_7$	$-10\omega_7$	EVII	3

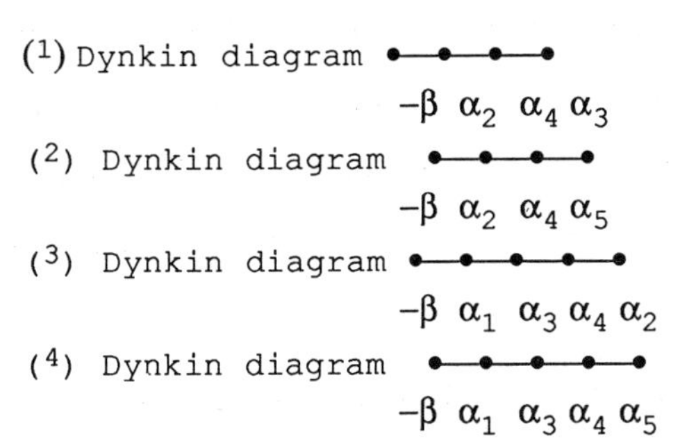

§7. The Oscillator Representation, Harmonic Polynomials and Associated Affine Varieties

In this section we consider the oscillator representation ω of the pair (G,G') for $G = SU(p,q)$, $Sp(n,\mathbb{R})$, and $G = SO^*(2m)$ and for G' a compact group dual to G in the sense of Roger Howe's theory of dual reductive pairs. In this setting (which we call the harmonic setting) we obtain modules for all reduction points of $SU(p,q)$ and $Sp(n,\mathbb{R})$. For $SO^*(2m)$ we obtain modules for a set of reduction points including all with level of reduction two or more. The precise description is given below as Proposition 7.16. This connection between unitarity and dual pairs is of course well known. It was an essential part of the articles [31] and [15] among many others. Some care is needed in describing the parameters for the components of the oscillator representations and we shall need all this detail in later sections. In providing this detail we point out an error found in the literature.

Also in this section we show that the realization of a module in the harmonic setting determines an affine variety V and the module can be recovered as the image in the coordinate ring of V of certain harmonic valued polynomials. This result is detailed in (7.6) and (7.7). In the scalar cases (i.e., for the parameters in what is called the Wallach set) and when G/K is tube type, this correspondence between modules and coordinate rings of $K_{\mathbb{C}}$-orbits in $\mathfrak{p}_+$ can be found in the Harris and Jakobsen article [19].

Various realizations of the oscillator representation appear in the literature. The description we give below is fairly uniform for all three groups. It is based on Fock space and follows most closely the construction found in [9]. Here, however, it will suffice to give just the infinitesmal action. In this setting the space of harmonic polynomials plays a key role. This space decomposes multiplicity free under the action of $K \times G'$ and the Howe duality correspondence describes the connection between the $\mathfrak{k}$ and $\mathfrak{g}'$ highest weights. This correspondence is the key to a generalization of Young products and a refinement of the factorization theorem which appears in section eight.

We provide details sufficient for the purpose of this paper and refer the reader to [9], [31], [23] and [24] for further information. The following table describes the conventions we adopt for each case. Let $M_{r,s}$ denote the space of $r \times s$ complex matrices.

TABLE 7.1

G	n	K	$\mathfrak{p}_+$	G'	$G'_{\mathbb{C}}$
SU(p,q)	n = p+q	S(U(p)×U(q))	$M_{p,q}$	U(k)	GL(k,ℂ)
Sp(n,ℝ)	n	U(n)	$\{T\in M_{n,n}:T=T^t\}$	O(k)	O(k,ℂ)
SO*(2m)	n = 2m	U(m)	$\{T\in M_{m,m}:T=-T^t\}$	Sp(k)	Sp(k,ℂ)

The groups given in (7.1) are defined as follows: $SU(p,q) = \{g\in SL(p+q,\mathbb{C}) : g\begin{pmatrix} I_p & 0 \\ 0 & -I_q\end{pmatrix} g^* = \begin{pmatrix} I_p & 0 \\ 0 & -I_q\end{pmatrix}\}$, $Sp(n,\mathbb{C}) = \{g\in GL(2n,\mathbb{C}) : g^t\begin{pmatrix} 0 & I_n \\ -I_n & 0\end{pmatrix} g = \begin{pmatrix} 0 & I_n \\ -I_n & 0\end{pmatrix}\}$, $Sp(n,\mathbb{R}) = Sp(n,\mathbb{C})\cap SU(n,n)$, and $Sp(k) = Sp(k,\mathbb{C}) \cap U(2k)$. We follow [31] and define $O(k,\mathbb{C})$ to be $\{g \in GL(k,\mathbb{C}) : g^tJg = J\}$ where J is the matrix $\begin{pmatrix} 0 & I_d \\ I_d & 0\end{pmatrix}$ if $k = 2d$ and is $\begin{pmatrix} 0 & I_d & 0 \\ I_d & 0 & 0 \\ 0 & 0 & 1\end{pmatrix}$ if $k = 2d+1$. We put $SO^*(2m) = SU(m,m) \cap O(2m,\mathbb{C})$.

Note that there exists a unitary symmetric matrix $J^{1/2}$ such that $(J^{1/2})^2 = J$. The group $O(k,\mathbb{C})$ and the group $\{g \in GL(k,\mathbb{C}) : g^t g = I_k\}$ are then conjugate by $J^{-1/2}$. Then $O(k)$ in (7.1) refers to the conjugate of the real orthogonal $k \times k$ matrices. We see that in the three cases G' is a subgroup of unitary matrices.

Let k and n be positive integers as given in Table 7.1. Let M be the space of matrices over $\mathbb{C}$ as given below in Table 7.2. Set $F = P(M)$ with inner product given in 2.1. Several formulas will be given below which depend on the following description of the variable $z \in M$.

TABLE 7.2

G	M	z
$SU(p,q)$	$M_{n,k}$	$\binom{x}{y}$, $x \in M_{p,k}$, $y \in M_{q,k}$
$Sp(n,\mathbb{R})$	$M_{n,k}$	(xy), $x,y \in M_{n,d}$ if $k = 2d$ (xyt), $x,y \in M_{n,d}$, $t \in M_{n,1}$ if $k = 2d+1$
$SO^*(2m)$	$M_{m,2k}$	(xy), $x,y \in M_{m,k}$

Let ω denote the oscillator representation of G on the closure of F. (For $G = Sp(n,\mathbb{R})$ one needs to go to a two-fold cover.) The oscillator representation is unitary and the space of K-finite vectors is F. We will let ω denote the infinitesimal action as well. Below we give the action of $\mathfrak{p}_-$, $\mathfrak{p}_+$ and K. In order to do this we first define the map $\theta : M \to \mathfrak{p}_+$ as follows:

SU(p,q): $\theta(z) = xy^t$,

Sp(n,ℝ): a) $\theta(z) = \frac{1}{2}zJz^t = \frac{1}{2}(yx^t + xy^t)$, if $k = 2d$,

b) $\theta(z) = \frac{1}{2}zJz^t = \frac{1}{2}(yx^t + xy^t + tt^t)$, $k = 2d+1$,

SO*(2m): $\theta(z) = \frac{1}{2}(xy^t - yx^t)$.

7.3 The action of $\mathfrak{p}_-$: Let $\varphi \in \mathfrak{p}_- = (\mathfrak{p}_+)^*$. Then $\omega(\varphi)$ is a multiplication operator given by $\omega(\varphi)f = (\varphi\circ\theta)f$, $f \in F$.

7.4 The action of $\mathfrak{p}_+$: Let $T \in \mathfrak{p}_+$. Then $\omega(T)$ acts by differentiation. More precisely, $\omega(T)f = -g(\partial)f$ where $g(z) = \mathrm{tr}(\theta(z)T^*)$, elements of $\mathfrak{p}_+$ are regarded here as matrices and $\theta(z)T^*$ is the matrix multiplication of $\theta(z)$ and the conjugate transpose of T. For those readers who would prefer a more invariant definition, $g(z)$ is a constant multiple (depending on $\mathfrak{g}_0$) of the map $z \to \langle\theta(z),\sigma T\rangle$ with $\langle\ ,\ \rangle$ the Killing form and σ the conjugate linear antiautomorphism defined in section one. If we fix a basis $\{T_i\}$ of $\mathfrak{p}_+$ then the corresponding functions $z \to g(z)$ will be called the *matrix entries* of $z \to \theta(z)$.

7.5 The action of K: This is given casewise.

SU(p,q): Let $u = (u_1,u_2) \in K$.

Then $\omega(u)f(z) = (\det u_1)^{-k} f\begin{pmatrix} u_1^{-1} & x \\ u_2^t & y \end{pmatrix}$

Sp(n,ℝ): The two fold cover K_2 of K appropriate for this case is $K_2 = \{(u,t) \in U(n) \times \mathbb{C}: \det u = t^2\}$ and group multiplication is coordinatewise. For $(u,t) \in K_2$, we have $\omega(u,t)f(z) = t^{-k}f(u^{-1}z)$.

SO*(2m): Let $u \in U(m) = K$. Then $\omega(u)f(z) = (\det u)^{-k} f(u^{-1}z)$.

The oscillator representation of $\mathfrak{g}$ on F decomposes into the direct sum of irreducible unitarizable highest weight modules (cf.(7.12),(7.13)). Thus the results of section five apply and in this setting give realizations on a coordinate ring of an affine algebraic variety. We take a slight detour from the main theme of this section to develop this point.

7.6 LEMMA. Let $\Theta: \mathbb{F} \to F$ be the map induced from the harmonic representation (ω, F) as in section five. Then for $G \in \mathbb{F}$

$$\Theta G(z) = G(\theta(z), z), \quad z \in M.$$

PROOF. It is enough to prove this for G a monomial. Let $G = \varphi f$ where $\varphi = \varphi_1 \ldots \varphi_m$, $\varphi_i \in \mathfrak{p}_- = (\mathfrak{p}_+)^*$ $i = 1,\ldots,m$ and $f \in F$. By the formula given in (7.3)

$$\Theta G(z) = \Theta(\varphi f)(z) = (\omega(\varphi)f)(z) = \varphi(\theta(z))f(z)$$

$$= G(\theta(z), z).$$

Suppose h is a $\mathfrak{g}$-highest weight vector in F and let L (resp. H) be the $\mathfrak{g}$-span (resp. $\mathfrak{k}$-span) of h. Then L (resp. H) is an irreducible $\mathfrak{g}$-module (resp. $\mathfrak{k}$-module). We let $Gr(\theta) = \{(\theta(z), z), z \in M\}$ denote the graph of θ. Lemmas (7.6) and (5.2) then imply the following theorem.

7.7 THEOREM. The module L is realized in the coordinate ring of the variety $Gr(\theta)$. Moreover this subspace is precisely the image of $\mathbb{H}$ in the coordinate ring of $Gr(\theta)$.

This result also leads to a connection with $K_{\mathbb{C}}$-orbits in $\mathfrak{p}_+$. However we defer this discussion to section twelve.

We now turn our attention to the description of harmonic polynomials. The compact group G' acts on F by the formulas:

SU(p,q) $\quad R(v)f(z) = f\begin{pmatrix} x & v \\ y & (v^t)^{-1} \end{pmatrix}$

Sp(n,$\mathbb{R}$) $\quad R(v)f(z) = f(zv)$

SO*(2m) $\quad R(v)f(z) = f(zv)$

for $v \in G'$. These actions extend holomorphically by the same formulas to the complexification $G'_{\mathbb{C}}$ of G'. Let I be the space of functions in F invariant under the action of G'. Then the matrix entries of $z \to \theta(z)$ together with the constant functions generate I. We let I be the ideal generated by the matrix entries of $z \to \theta(z)$ and let H be the orthogonal complement of I. We call H the ***space of harmonic polynomials*** or merely the ***harmonics***. An element in H is called harmonic. This terminology is suggested in part by the following proposition.

7.8 PROPOSITION. Let $h \in F$. The following are equivalent:

1) h is harmonic

2) $g(\partial)h = 0$ for all g, a matrix entry of $z \to \theta(z)$

3) $\omega(x)h = 0$ for all $x \in \mathfrak{p}_+$.

PROOF. The equation $(g(\partial)h|f) = (h|gf)$ given in (2.8) establishes the equivalence of (1) and (2). The equivalence of (2) and (3) follows from (7.4).

The following corollaries, which have specific uses later in the paper, illustrate how plentiful harmonic polynomials are.

7.9 COROLLARY. Let $G = SU(p,q)$ and let r and s be integers such that $0 \leq r \leq p$, $0 \leq s \leq q$, and $r + s \leq k$. Let $A = A(z)$ denote the first r columns of x and $B = B(z)$ the last s columns of y. Let $V \subset F$ be the space of polynomials in the matrix entries of A and B. Then V is a $\mathfrak{k}$-submodule of H and a subalgebra of F.

PROOF. Let g be the (i,j) matrix entry of $z \to \theta(z)$. Then $g(\partial) = \sum_{v=1}^{k} \frac{\partial^2}{\partial x_{iv}\partial y_{jv}}$. Let $f \in V$. Since $r + s \leq k$ we have $\frac{\partial^2}{\partial x_{iv}\partial y_{jv}} f = 0$, for each v. Thus $g(\partial)f = 0$. By (7.8), $f \in H$. The other points are clear.

7.10 COROLLARY. Let $G = Sp(n,\mathbb{R})$. Let j be an integer such that $0 \leq j \leq \min(n,d)$. Let $A_1 = A_1(z)$ denote the first j columns of z and let $V_1 \subset F$ be the space of polynomials in the matrix entries of A_1. Then V_1 is a $\mathfrak{k}$-submodule of H and a subalgebra of F. Now let $\hat{y}$ denote the last $d-j$ columns of y and let $A_2 = A_2(z)$ denote the $n \times (k-j)$ matrix $(x\hat{y})$ if $k = 2d$ and $(x\hat{y}t)$ if $k = 2d + 1$. Suppose $k - j \leq n$. Let W be the linear span of the $(k-j)$-minors in the matrix A_2 and let $V_2 = V_1W$. Then V_2 is a $\mathfrak{k}$-submodule and V_1-submodule of H.

PROOF. By considering the matrix entries of $z \to \theta(z)$ and applying (7.8) we prove that $V_1 \subset H$ as in the previous corollary. The proof of Proposition 6.6 [p.23;31] may easily be adapted to prove $V_2 \subset H$.

7.11 COROLLARY. Let $G = SO^*(2m)$. Let V be the space of polynomials in the matrix entries of x. Then V is a $\mathfrak{k}$-submodule of H and a subalgebra of F.

PROOF. Again this follows by considering matrix entries of $z \to \theta(z)$ and applying (7.8).

The following proposition reduces the $\mathfrak{g}$-decomposition of F to the K-decomposition of H.

7.12 PROPOSITION. The space of harmonics H is invariant under the action of $K \times G'$. Let V be a $\mathfrak{k}$-invariant (resp. irreducible) space of harmonics. Then IV is a $\mathfrak{g}$-module (resp. irreducible highest weight module). If $V = \oplus V_i$ is an orthogonal $\mathfrak{k}$-decomposition then $IV = \oplus IV_i$ is an orthogonal $\mathfrak{g}$-decomposition. Furthermore $F = IH$.

PROOF. The action of $K \times G'$ is unitary and *I* is easily seen to be invariant. The $K \times G'$ invariance of H therefore follows. Suppose V is a $\mathfrak{k}$-invariant space in H. By (7.3) and (7.5) it is clear that IV is invariant under $\mathfrak{p}_- \oplus \mathfrak{k}$. By (7.8) $\omega(\mathfrak{p}_+)V = 0$ and since $IV = \omega(U(\mathfrak{p}_-))V$, $[\mathfrak{k},\mathfrak{p}_-] \subset \mathfrak{p}_-$, and $[\mathfrak{p}_+,\mathfrak{p}_-] \subset \mathfrak{k}$ it follows that IV is $\omega(\mathfrak{p}_+)$ invariant. Thus IV is a $\mathfrak{g}$-module. Now suppose V is irreducible. Let I_1 be the span of the matrix entries of $z \to \theta(z)$. Set $I_m = (I_1)^m$ for $m \geq 1$ and $I_0 = \mathbb{C}$. Then IV is the direct sum of the $\mathfrak{k}$-subspaces I_mV, $m \geq 0$. If S is a nonzero submodule of IV, then there exists a least integer m_0 such that $S \cap I_{m_0}V \neq 0$. Let ϕ be a nonzero vector in $S \cap I_{m_0}V$. Since $\omega(\mathfrak{p}_+)I_{m_0}V \subset I_{m_0-1}V$, we have $\omega(\mathfrak{p}_+)\phi = 0$. By (7.8), ϕ is harmonic. This implies $m_0 = 0$. By the irreducibility of V we have $V = S \cap V$. Hence $S = IV$. It follows that IV is irreducible. By (7.8)

$\mathfrak{p}_+$ annihilates V. This implies IV is a highest weight module. The unitarity of ω implies that any orthogonal $\mathfrak{k}$-decomposition of V induces an orthogonal $\mathfrak{g}$-decomposition of IV. Finally, by (7.4) F is the direct sum of simple highest weight modules. Moreover by (7.8) all highest weights are harmonic polynomials. So $F = I \cdot H$.

7.13 The Howe duality correspondence. The space of harmonics decomposes into a multiplicity free orthogonal direct sum of irreducible $K \times G'$ modules $H(\lambda \otimes \sigma)$. By (7.12) this induces a multiplicity free orthogonal decomposition of F into irreducible $\mathfrak{g} \times G'$ modules $F(\lambda \otimes \sigma) = IH(\lambda \otimes \sigma)$, where λ and σ are the $\mathfrak{g}$ and $\mathfrak{g}'$-highest weights, respectively. The weight λ is uniquely determined by σ. We will thus write $\lambda = \lambda(\sigma)$, $H^\sigma = H(\lambda(\sigma) \otimes \sigma)$ and $F^\sigma = F(\lambda(\sigma) \otimes \sigma)$. Let $\Sigma(k)$ be the set of $\mathfrak{g}'$-highest weights σ such that H^σ is nonzero in H and let $\Lambda(k) = \lambda(\Sigma(k))$. The map $\lambda\colon \Sigma(k) \to \Lambda(k)$ is a bijection and is known as the duality correspondence. We have the multiplicity free decompositions $H = \bigoplus_{\sigma \in \Sigma(k)} H^\sigma$ and $F = \bigoplus_{\sigma \in \Sigma(k)} F^\sigma$.

7.14 The cone decomposition and Fock space

In section six we introduced the cone decomposition of Λ_r. Let k and $\Lambda(k)$ be as above and let $\Lambda_r(k)$ be the set of reduction points in $\Lambda(k)$. We shall call $\Lambda_r(k)$ the Fock space parameters (for k). For convenience we set $\Lambda_r(0) = \{0\} \subset \mathfrak{h}^*$. We now describe the relationship between the cones Λ_a and the Fock space parameters $\Lambda_r(k)$.

7.15 DEFINITIONS. (1) Let the constant C be given in Table (6.20). Define the linear functional $\underline{k}$ on $\mathfrak{h}^*$ by the formula:

$$\underline{k}(\lambda) = -\frac{2(\lambda,\beta)}{C(\beta,\beta)}$$

where as usual β is the maximal root of Δ^+.

(2) Let $\Omega = \cup_{k=0}^{\infty} \Lambda_r(k)$ and note that for large k the $\Lambda_r(k)$ are empty.

(3) For the case $G = SO^*(2m)$ let E denote the hyperplane in $\mathfrak{h}^*$ orthogonal to the simple root $\alpha_1 = e_1 - e_2$.

In the classification article [15] the introductory remarks include the assertion (in somewhat different language) that the sets Ω and Λ_r are equal. This is in fact not true. The following proposition describes the precise relationship. For G equal to $SU(p,q)$ or the double cover of $Sp(n,\mathbb{R})$ these sets are equal. This was conjectured by Kashiwara and Vergne [31] and established in [14].

7.16 PROPOSITION. (1) The restriction of $\underline{k}(.)$ to each cone Λ_a, $a \in \mathcal{A}$, is constant. Let k_a denote this constant value. In addition, for each $k \in \mathbb{N}$ the restriction of $\underline{k}(.)$ to $\Lambda_r(k)$ is constant with value k, provided of course that $\Lambda_r(k) \neq \emptyset$. So $\Lambda_r(s) \cap \Lambda_r(t)$ is empty for $s \neq t$.

(2) Suppose $G = SU(p,q)$ or $Sp(n,\mathbb{R})$. Then $\Lambda_r(k)$ is the disjoint union of all the cones Λ_a where $k_a = k$.

(3) Suppose $G = SO^*(2m)$. Then $\Lambda_r(k)$ is the disjoint union of all the sets $\Lambda_a \cap E$ where $k_a = k$.

7.17 REMARK. The remainder of this section is devoted to the proof of Proposition (7.16). This is accomplished in a case by case fashion by giving an explicit description of the sets $\Lambda_r(k)$. An immediate consequence of (7.16) and the cone decomposition (6.6) is the following corollary.

7.18 COROLLARY.

(1) Ω is the disjoint union of $\Lambda_r(k)$, $k \in \mathbb{N}$

(2) $\Omega = \Lambda_r$ for $G = SU(p,q)$ or $Sp(n,\mathbb{R})$.

(3) $\Omega = \Lambda_r \cap E$ for $G = SO^*(2m)$.

7.19 REMARK. We note that in all cases we have $\Lambda_a \cap \Omega \subset \Lambda_r(k_a)$.

We now turn our attention to the description of $\Lambda(k)$.

The group $G'_{\mathbb{C}}$ is one of the complex classical Lie groups $GL(k,\mathbb{C})$, $O(k,\mathbb{C})$ or $Sp(k,\mathbb{C})$. We let $\mathfrak{h}'$ and $\mathfrak{b}'$ denote the standard Cartan and Borel subalgebras, respectively, of the Lie algebra $\mathfrak{g}'$ of $G'_{\mathbb{C}}$. More specifically, $\mathfrak{h}'$ is the subalgebra of diagonal elements of $\mathfrak{g}'$. For $gl(k,\mathbb{C})$, $\mathfrak{b}'$ is the subalgebra of upper triangular matrices; for $o(k,\mathbb{C})$, $\mathfrak{b}'$ is the set of matrices of the form

$$\text{(a)} \quad \begin{pmatrix} A & B \\ 0 & -A^t \end{pmatrix} \qquad \text{k-even,}$$

$$\text{(b)} \quad \begin{pmatrix} A & B & C \\ 0 & -A^t & 0 \\ 0 & -C^t & 0 \end{pmatrix} \qquad \text{k-odd,}$$

where A is upper triangular and B is skew symmetric; for $Sp(k,\mathbb{C})$, $\mathfrak{b}'$ is the set of matrices of the form $\begin{pmatrix} A & B \\ 0 & -A^t \end{pmatrix}$ where A is upper triangular and B is symmetric.

For the Lie algebra $\mathfrak{g}$ we choose Cartan and Borel subalgebras $\mathfrak{h}$ and $\mathfrak{b}$, respectively, compatible with the choices of positive roots as described in sections seven, eight, and nine of [15]. More specifically, let $\mathfrak{b}(n)$ denote the upper triangular $n \times n$ matrices over $\mathbb{C}$. Let $\mathfrak{b}_k$ denote the Borel subalgebra of $\mathfrak{k}$ given by:

$SU(p,q)$: $\mathfrak{b}_k = \{(a,b) \in \mathfrak{b}(p) \times \mathfrak{b}(q) : Tr(a) + Tr(b) = 0\}$,

$Sp(n,\mathbb{R})$: $\mathfrak{b}_k = \mathfrak{b}(n)$,

$SO^*(2m)$: $\mathfrak{b}_k = \mathfrak{b}(m)$.

We put $\mathfrak{b} = \mathfrak{b}_k \oplus \mathfrak{p}_+$ and let $\mathfrak{h}$ be the diagonal matrices in $\mathfrak{b}_k$.

With these choices of Borel and Cartan subalgebras we now proceed to give the $\mathfrak{g} \times \mathfrak{g}'$-highest weight vectors of $F = \bigoplus_{\sigma \in \Sigma(k)} F^\sigma$ in each of the three cases. To do this we adopt the convention that whenever x is a matrix, then $\Delta_i(x)$ will denote the $i \times i$ lower left minor of x and $\varepsilon_j(x)$ will denote the upper right $j \times j$ minor of x. We also set $\Delta_0(x) = \varepsilon_0(x) = 1$.

7.20 SU(p,q)×U(k)-highest weight vectors. Fix integers r and s such that $0 \le r \le p$, $0 \le s \le q$, and $r + s \le k$. Let $(a_1,\ldots,a_r)$ and $(b_1,\ldots,b_s)$ be sequences of nonnegative integers and let

$$\phi(z) = \Delta_1^{a_1}(x) \ldots \Delta_r^{a_r}(x)\, \varepsilon_1^{b_1}(y) \ldots \varepsilon_s^{b_s}(y).$$

By (7.9), we see $\phi \in H$. It is easy to check that ϕ is a $\mathfrak{g}\times\mathfrak{g}'$-highest weight vector. Furthermore, all $\mathfrak{g}\times\mathfrak{g}'$-highest weight vectors are of this form (cf. Prop. 6.1 [31]). The $\mathfrak{g}$-highest weight for ϕ is

$$\sum_{j=1}^{p-1} a_j\omega_{p-j} - (\sum_{j=1}^{p} a_j + k + \sum_{j=1}^{q} b_j)\omega_p + \sum_{j=1}^{q-1} b_j\omega_{j+p}$$

where $a_i = 0$ for $i > r$ and $b_i = 0$ for $i > s$.

7.21 $Sp(n,\mathbb{R}) \times O(k)$-highest weight vectors. Fix an integer j such that $0 \leq j \leq \min(d,n)$. Let $(a_1,\ldots,a_j)$ be a sequence of nonnegative integers. Let

$$(1)\quad \phi(z) = \Delta_1^{a_1}(x)\ \ldots\ \Delta_j^{a_j}(x).$$

Now suppose $k - j \leq n$. Let a be the lower left $(k-j) \times d$ submatrix of z and let b be the lower right $(k-j) \times (k-j-d)$ submatrix of z. Then (ab) is a $(k-j) \times (k-j)$ matrix. Let $\tilde{\Delta}_j(z) = \det(ab)$ and let

$$(2)\quad \phi(z) = \Delta_1^{a_1}(x)\ \ldots\ \Delta_j^{a_j}(x)\tilde{\Delta}_j(z)\ .$$

By (7.10), we know $\phi \in H$ and it is straightforward to verify ϕ is a $\mathfrak{g}\times\mathfrak{g}'$-highest weight vector. All $\mathfrak{g}\times\mathfrak{g}'$-highest weight vectors are of this form (cf. Prop. 6.6 [31]). The $\mathfrak{g}$-weight of ϕ, is:

$$(3)\quad a_{n-1}\omega_1 + \ldots + a_1\omega_{n-1} - \left(a_1 + \ldots + a_n + \frac{k}{2}\right)\omega_n$$

where $a_i = 0$ for $i > j$

$$(4)\quad a_{n-1}\omega_1+\ldots+a_1\omega_{n-1} + \omega_{n-k+j} - \left(1 + a_1 + \ldots + a_n + \frac{k}{2}\right)\omega_n$$

where $a_i = 0$ for $i > j$.

Note that when $n-k+j = 0$, by convention the weight ω_{n-k+j} in (4) doesn't appear. Also note that some weights can be expressed in either form (3) or (4) above. Of course when this occurs the two expressions (1) and (2) for $\phi(z)$ will also be equal.

7.22 $SO^*(2m) \times Sp(k)$-highest weight vectors. Let r be an integer with $0 \leq r \leq \min(m,k)$. Let $(a_1,\ldots,a_r)$ be a sequence of nonnegative integers. Let

$$\phi(z) = \Delta_1^{a_1}(x) \ldots \Delta_r^{a_r}(x).$$

Then by (7.11) $\phi \in H$ and it is easy to verify that ϕ is a $\mathfrak{g}\times\mathfrak{g}'$-highest weight vector. The proof of Proposition 6.1 [31] may be adapted to show that all $\mathfrak{g}\times\mathfrak{g}'$-highest weight vectors are of this form. The $\mathfrak{g}$-highest weight of ϕ is

$$a_{m-1}\omega_1 + \ldots + a_1\omega_{m-1} - (2k+a_1+2(a_2 + \ldots + a_m))\omega_m$$

where $a_i = 0$ for $i > r$.

Proof of (7.16). Let $a = (Q,R,\ell) \in \mathcal{A}$. We recall that C_a is orthogonal to β. Since $\Lambda_a = \lambda_a + C_a$ by (6.6), it follows that $\underline{k}$ is constant on Λ_a. Let k_a denote this value of $\underline{k}$ on Λ_a. We now consider cases.

$SU(p,q)$ We begin by showing $\underline{k}$ has value k on $\Lambda_r(k)$. Since $\Lambda_r(0) = \{0\}$, the case $k = 0$ is clear so we assume $k > 0$. Let $\lambda \in \Lambda(k)$. Then $\lambda \in \Lambda$ so that $Q(\lambda) = R(\lambda) = SU(p',q')$ where $1 \leq p' \leq p$ and $1 \leq q' \leq q$. By the explicit form of λ given in (7.20) we have $\underline{k}(\lambda) = k+a_p+b_q$. So, it suffices to show that if $a_p > 0$ or $b_q > 0$, then $\lambda \notin \Lambda_r$. By the construction of the highest weight vector ϕ in (7.20) we have $k \geq r + s$. Now $a_p > 0$ implies $r = p$ and $\lambda \in \Lambda_a$ implies $b_{q-q'} \neq 0$ which in turn gives $s \geq q - q'$. So $k \geq n - q'$. Write $\lambda = \lambda_0 + z\zeta$ as in section six. We now show $z < A$ where A is the first reduction point as described in (6.1) and $\lambda = \lambda(z)$. From Lemma 7.3 [15] we have $A = \max\{p',q'\}$. Since $z = n-k-1-(a_p+b_q)$ the constraint $k \geq p+(q-q')$ implies $z < A$. Thus we have $\lambda \notin \Lambda_r$. If $b_q > 0$ then the constraint $k \geq q+(p-p')$ again implies $z < A$. Consequently we have $\underline{k}(\lambda) = k$ if $\lambda \in \Lambda_r(k)$.

Let λ_a be given as in Table (6.22). Then $k_a = n+\ell-p'-q'-1$ is a nonnegative integer. By the explicit form of λ_a and C_a (cf. (6.21)) one sees that $\lambda \in \Lambda_a = \lambda_a + C_a$ is of the form given in (7.20) if one takes k to be k_a. This shows $\Lambda_a \subset \Lambda_r(k_a)$. Furthermore, $\Lambda_a \cap \Lambda_r(k)$ is empty for $k \neq k_a$ since $\underline{k}$ takes the value k on $\Lambda_r(k)$. Now put $A = \{a \in \mathcal{A}: k_a = k\}$. Then from the cone decomposition we obtain the disjoint decomposition:

$$\Lambda_r(k) = \bigcup_{a \in A} \Lambda_a .$$

Sp(n, ℝ) The case $k = 0$ being clear, we let $\lambda \in \Lambda(k)$ where $k > 0$. Since $\lambda \in \Lambda$ we have $Q(\lambda) = Sp(q,\mathbb{R})$ and $R(\lambda) = Sp(r,\mathbb{R})$ where $1 \leq q \leq r \leq n$. By the form of λ in (7.21) one finds $\underline{k}(\lambda) = k+2a_n$. Writing $\lambda = \lambda_0+z\zeta$ as in section six, we find $z = n-a_n-k/2$. If $a_n > 0$ then by the construction of the highest weight vector ϕ in (7.21) we have $d \geq n$ where d is $k/2$ if k is even and is $(k-1)/2$ if k is odd. Thus $a_n > 0$ implies $z < 0$ so that $\lambda \notin \Lambda_r$. Indeed, since $z < 0$, the weight λ corresponds to a discrete series by Theorem 2.4 [15]. This shows that $\underline{k}$ has value k on $\Lambda_r(k)$.

Let λ_a be given as in Table (6.22). Then $k_a = 2n-(r+q)+\ell-1$ is a nonnegative integer. As in the SU(p,q) case one uses (6.21) to compare the form of $\lambda \in \Lambda_a = \lambda_a + C_a$ to the form given in (7.21) to see that $\lambda \in \Lambda_r(k)$ for $k = k_a$. The remainder of the argument proceeds exactly as in the SU(p,q) case.

SO*(2m) Suppose $\lambda \in \Lambda(k)$ for some $k > 0$ and λ is given by formula (7.22). Then $\underline{k}(\lambda) = \frac{1}{2}(a_{m-1} + 2(k + a_m))$. Suppose either a_{m-1} or a_m is nonzero. Then by the construction of the highest weight vector ϕ in (7.22) we have $k \geq r \geq m-1$. Writing $\lambda = \lambda_0 + z\zeta$ as in section six, a straightforward calculation gives

$$z = -2k + 2m - 3 - a_{m-1} - 2a_m \leq -2(m-1) + 2m - 3 = -1.$$

This implies λ is a highest weight for a discrete series representation by Theorem 2.4 of [15]. In particular $\lambda \notin \Lambda_r$. This shows that if $\lambda \in \Lambda_r(k)$ then $k < m-1$ and so $(\lambda,\alpha_1) = a_{m-1} = a_m = 0$. Thus $\underline{k}(\lambda) = k$ if $\lambda \in \Lambda_r(k)$ and $\Lambda_r(k) \subset E$.

To prove (3) we let $\lambda \in \Lambda_a$, $a \in \mathcal{A}$. If $\lambda \in \Lambda_r(k)$ for some $k \in \mathbb{N}$ then clearly $k_a = \underline{k}(\lambda) = k$ and $(\lambda,\alpha_1) = a_{m-1} = 0$. On the other hand suppose $(\lambda,\alpha_1) = 0$. This implies by (6.21) and (6.22) that $Q(\lambda) = SO^*(2p)$, $3 \leq p \leq m$, or $Q(\lambda) = SU(1,1)$. In these cases, $k_a = \ell-1+m-p$ or $k_a = m-2$, respectively. We see k_a is a nonnegative integer. Furthermore λ is of the form given by (7.22) for $k = k_a$. Thus we have $\lambda \in \Lambda_r(k_a)$. Since $\underline{k}$ has value k on $\Lambda_r(k)$, $\Lambda_a \cap E \cap \Lambda_r(k)$ is empty for $k \neq k_a$. Now let $A = \{a \in \mathcal{A}: k_a = k\}$. From the cone decomposition (6.6) we obtain the disjoint decomposition

$$\Lambda_r(k) = \cup_{a \in A} \Lambda_a \cap E .$$

This completes the proof of (7.16).

7.23 For later calculations it is convenient to relate the formulas (7.21)(3) and (4) with those in the tables of section six. In particular we indicate which vertices have weights of the form (3) and (4) respectively.

Let $G = Sp(n,\mathbb{R})$. Suppose $a = (Q,R,\ell) \in \mathcal{A}$. Then $Q = Sp(q,\mathbb{R})$ and $R = Sp(r,\mathbb{R})$, $1 \leq q \leq r \leq n$. Let $k = k_a = 2n - (r+q-\ell+1)$. By (6.22) we have $\lambda_a = \omega_q + \omega_r - \left(2 + \frac{k}{2}\right)\omega_n$ and by (7.16) $\lambda_a \in \Lambda_r(k)$. We now partition $\mathcal{A}$ into two disjoint sets $\mathcal{A}_1$ and $\mathcal{A}_2$, where $a \in \mathcal{A}_1$ if $2 \leq \ell-(r-q)$ and $a \in \mathcal{A}_2$ if $1 \geq \ell-(r-q)$. These conditions imply that for $a \in \mathcal{A}_1$, $n-r \leq n-q \leq d$ and for $a \in \mathcal{A}_2$, $n-r \leq d$ and $k-n+q \leq d$. From (7.21) if ϕ_a denotes the highest weight vector with weight λ_a then

(1) $\phi_a(z) = \Delta_{n-r}(x)\Delta_{n-q}(x)$ if $a \in \mathcal{A}_1$

(2) $\phi_a(z) = \Delta_{n-r}(x)\tilde{\Delta}_{k-n+q}(z)$ if $a \in \mathcal{A}_2$.

7.24 REMARK. For later reference we note that for $\lambda \in \Omega$ the expressions for λ given in (7.20), (7.21) and (7.22) satisfy $a_p = b_q = 0$, $a_n = 0$ and $a_{m-1} = a_m = 0$ in the three cases respectively.

We also record the values of k_a given above in the proof of (7.16).

TABLE 7.25

G	k_a
$SU(p,q)$	$n + \ell - p' - q' - 1$
$Sp(n,\mathbb{R})$	$2n - r - q + \ell - 1$
$SO^*(2m)$	$\ell - 1 + m - p$ (let $p=2$ for $Q \sim SU(1,1)$)

§8. Young Products and a Refinement of the Factorization Theorem

In this section we begin with a generalization and application of the theory of Young products to the setting of harmonic polynomials. Following this we recall the notation surrounding the Factorization Theorem (6.15) and observe that in our current setting the $\mathfrak{k}$-module map P can be given explicitly as multiplication of polynomials. This implies that the map P is also given as multiplication of polynomials and so we obtain a more precise connection between the $\mathfrak{k}$-module J°_{λ} and the $\mathfrak{k}$-module for the vertex, J°_{a}. This connection is given as a Young product; i.e. for $\lambda \in \Lambda_a$, $J^{\circ}_{\lambda} = J^{\circ}_{a} H'$ with juxtaposition designating multiplication of polynomials. This specialized reformation of (6.15) is given below as (8.5).

We retain the notation of the previous section and now turn to a generalization of the theory of Young products. For λ the $\mathfrak{g}$-highest weight of the $\mathfrak{g} \times \mathfrak{g}'$ highest weight vector ϕ given in (7.20), (7.21) and (7.22) we let H_λ be the irreducible K-space generated by ϕ.

8.1 PROPOSITION. Let $\sigma_1, \sigma_2 \in \Sigma(k)$ and $\lambda_i = \lambda(\sigma_i)$, $i = 1,2$. Suppose $H_{\lambda_1} H_{\lambda_2} \subset H$. Then $\sigma_1 + \sigma_2 \in \Sigma(k)$ and $H_{\lambda_1} H_{\lambda_2} = H_{\lambda(\sigma_1+\sigma_2)}$.

PROOF. The Borel subalgebra $\mathfrak{b}'$ of $\mathfrak{g}'$ acts on every function in $H_{\lambda_1} H_{\lambda_2}$ by the weight $\sigma_1 + \sigma_2$. Since $H_{\lambda_1} H_{\lambda_2} \subset H$, we have

$\sigma_1 + \sigma_2 \in \Sigma(k)$. Furthermore any K-highest weight vector ϕ in $H_{\lambda_1} H_{\lambda_2}$ is a highest weight vector for $K \times G'$. By the duality correspondence ϕ has highest weight $\lambda(\sigma_1+\sigma_2)\otimes(\sigma_1+\sigma_2)$. Since the $K\times G'$-decomposition of H is multiplicity free (7.13), ϕ is unique up to a scalar. Hence $H_{\lambda_1}H_{\lambda_2} = H_{\lambda(\sigma_1+\sigma_2)}$.

8.2 REMARKS. (1) It is worthy of note that $\lambda(\sigma_1 + \sigma_2) \neq \lambda(\sigma_1) + \lambda(\sigma_2)$. Recall from Table 6.20 that $C = 1, \frac{1}{2}$, or 2 according as G is $SU(p,q)$, $Sp(n,\mathbb{R})$ of $SO^*(2m)$. Let ω_0 be the fundamental dominant weight corresponding to the unique noncompact simple root. Then

$$\lambda(\sigma_1 + \sigma_2) = \lambda(\sigma_1) + \lambda(\sigma_2) + Ck\omega_0.$$

(2) Evidently the proof of the above proposition remains valid for other choices of Borel subalgebra $\mathfrak{b}\times\mathfrak{b}'$ of $\mathfrak{g}\times\mathfrak{g}'$ provided that one appropriately chooses H_λ, $\lambda = \lambda(\sigma)$, as the K-invariant space spanned by the $\mathfrak{b}\times\mathfrak{b}'$-highest weight vector in H^σ.

(3) Proposition (8.1) generalizes the classical notion of Young products (cf. [p. 278, 39] or [p. 136, 47]). To see this let $G = SO^*(2m)$. Then functions in $H_{\lambda_1}H_{\lambda_2}$ are independent of the variable y. This implies by (7.11) that $H_{\lambda_1}H_{\lambda_2} \subset H$. The proposition then yields the classical result.

8.3 LEMMA. Let $\lambda \in \Omega$ and $a = a(\lambda)$. Write $\lambda = \lambda_a + \lambda'$, where λ_a is the vertex of the cone Λ_a and $\lambda' \in C_a$. Let $k = k_a$ be as in (7.25) and let σ and σ_a be elements of $\Sigma(k)$ such that $\lambda = \lambda(\sigma)$ and $\lambda_a = \lambda(\sigma_a)$. If $\sigma' = \sigma - \sigma_a$. Then $\sigma' \in \Sigma(k)$ and $\lambda' = \lambda(\sigma') + kC\omega_0$.

PROOF. Let $\lambda \in \Omega$ and $a = a(\lambda)$. Write $\lambda = \lambda_a + \lambda'$. It is straightforward using (6.21) and (7.20), (7.21) and (7.22) to verify that $\lambda' - kC\omega_0$ is an element of $\Lambda(k)$ if $k = k_a$. By the duality correspondence there exists σ, σ_a, and $\sigma' \in \Sigma(k)$ such that $\lambda = \lambda(\sigma)$, $\lambda_a = \lambda(\sigma_a)$, and $\lambda' - kC\omega_0 = \lambda(\sigma')$. But then

$\lambda(\sigma) = \lambda(\sigma_a) + \lambda(\sigma') + kC\omega_o = \lambda(\sigma_a + \sigma')$ by (8.2.1). Since the function $\sigma \to \lambda(\sigma)$ is a bijection we have $\sigma = \sigma_a + \sigma'$. Thus $\sigma' = \sigma - \sigma_a$.

With the notation as in (8.3) we let $H' = H_{\lambda(\sigma')} \otimes \mathbb{C}_{kC\omega_o}$. As a vector space we identify H' with $H_{\lambda(\sigma')}$. By (8.3) the highest weight of H' is $\lambda(\sigma') + kC\omega_o = \lambda'$. For convenience we set $H_a = H_{\lambda_a}$.

8.4 PROPOSITION. Let $\lambda \in \Omega$, $a = a(\lambda)$ and $\lambda = \lambda_a + \lambda'$. The map $P: H_a \otimes H' \to F$ given by $P(f \otimes g) = fg$ is a K-intertwining map whose image is H_λ.

PROOF. It is easy to see that P intertwines the K-actions. By (8.1) and (8.2) it is sufficient to verify that $H_aH' \subset H$. We verify this casewise. Suppose $G = SU(p,q)$ and $Q = R = SU(p',q')$. Let $r = p - p'$ and $s = q - q'$. Then $k = \ell - 1 + r + s$ and since $\ell \geq 1$ we have $r + s \leq k$. Let V be the $\mathfrak{k}$-module and algebra of functions defined as in (7.9). By (7.20) the highest weight vector of H_a is $z \to \Delta_r(x)\varepsilon_s(y)$ while the highest weight vector of H' is of the form $z \to \Delta_1^{a_1}(x)\ldots\Delta_r^{a_r}(x)\ \varepsilon_1^{b_1}(y)\ldots\varepsilon_s^{b_s}(y)$; both are in V. Since V is an algebra H_a, H' and H_aH' are in V. So (7.9) implies $H_aH' \subset H$ establishing the proposition in this case.

Now suppose $G = Sp(n,\mathbb{R})$, $Q = Sp(q,\mathbb{R})$ and $R = Sp(r,\mathbb{R})$. Then $k = 2n - (r+q) + \ell - 1$. We consider two subcases. Suppose $2 \leq \ell - (r-q)$. Then $n-q \leq d$. Let V_1 be the algebra of functions defined as in (7.10) with $j = n-q$. By (7.21) the highest weight vector of H_a is $z \to \Delta_{n-r}(x)\ \Delta_{n-q}(x)$ and the highest weight vector of H' is of the form $\Delta_1^{a_1}(x)\ldots\Delta_{n-r}^{a_r}(x)$; both are in V_1. Since V_1 is an algebra H_a, H', and H_aH' are in V_1. So (7.10) implies $H_aH' \subset H$. Now suppose

$1 \geq \ell - (r-q)$. Then $k - n + q \leq d$. Let V_1 be the algebra and W be the V_1-module defined in (7.10) with $j = k - n + q$. By (7.21) the highest weight vector of H_a is $z \to \Delta_{n-r}(x)\ \tilde{\Delta}_{k-n+q}(z)$ and the highest weight vector of H' is of the form $z \to \Delta_1^{a_1}(x)\ldots\Delta_{n-r}^{a_{n-r}}(x)$. Since $n - r \leq j$ the highest weight vector of H_a is in V_1W and the highest weight vector of $H' \in V_1$. So (7.10) implies that $H_aH' \subset H$ establishing the proposition in this case.

Now suppose $G = SO^*(2m)$. Let V be the algebra of functions defined in (7.11). By (7.22) any highest weight vector is in V. Hence H_a, H' and H_aH' are in V. Here (7.11) establishes the proposition in this case.

Applying the preceding proposition yields the following refinement of the Factorization Theorem (6.15).

8.5 COROLLARY. Let $\lambda \in \Omega$, and $a = a(\lambda)$. Then $\overset{\circ}{J}_\lambda = \overset{\circ}{J}_a H'$ and $J_\lambda = J_a H' = P(\mathfrak{p}_+,\mathbb{C})\overset{\circ}{J}_a H'$.

8.6 REMARK. Note that the refinement of the factorization theorem given above may be deduced directly from (5.2), (7.6) and (8.4). More specifically, (8.4) implies that $\overset{\circ}{J}_a\ H'$ is a $\mathfrak{k}$-invariant subspace of $\mathbb{H}_\lambda$ and consists of polynomials of degree $\ell(\lambda_a) = \ell(\lambda)$. If $F = Gh$, $G \in \overset{\circ}{J}_a$ and $h \in H'$, then by (7.6) $\Theta F(z) = \Theta G(z)\cdot h(z)$, $z \in M$. By (5.2) G and thus F are in the kernel of Θ. Hence $\overset{\circ}{J}_a\ H' \subset \overset{\circ}{J}_\lambda$.

§9. The Fundamental System of Differential Operators

In the harmonic setting the factorization theorem (8.5) demonstrates the ubiquitous role of the vertex K-module J_a^o. It appears as a factor in J_λ^o for each λ in the cone Λ_a. This role for J_a^o leads to a finite system $\mathcal{F}_a$ of differential operators defined through the pairing (2.2). Unlike the system $\mathcal{D}_a$, which acts on a unique polynomial space, the system $\mathcal{F}_a$ acts on $\mathbb{H}_\lambda$ for all λ in the cone Λ_a. Here we prove that the kernel of $\mathcal{F}_a$ in $\mathbb{H}_\lambda$ is precisely the irreducible submodule $L(\lambda+\rho)$. Since the number of cones is finite we obtain finitely many systems $\mathcal{F}_a$, $a \in \mathcal{A}$, which determine all the unitarizable modules.

Throughout this section we shall view the space $\mathbb{F} = P(\mathfrak{p}_+, F)$ as the space of scalar-valued polynomials on $\mathfrak{p}_+ \times M$. By (2.2) differential operators constructed from such polynomials will now include differentiation in the variable $z \in M$. To underscore this distinction we let $p(\partial_o): P(\mathfrak{p}_+ \times M, \mathbb{C}) \to P(\mathfrak{p}_+ \times M, \mathbb{C})$ denote this differential operator in both the $\mathfrak{p}_+$ and M variables.

9.1 DEFINITION. Let $a \in \mathcal{A}$ and let $m_1, \ldots, m_n$ be a (linear) basis for J_a^o. The set $\mathcal{F}_a = \{m_1(\partial_o), \ldots, m_n(\partial_o)\}$ will be referred to as a ***fundamental system (associated to the cone*** Λ_a***)***.

As in section eight, for $\lambda \in \Omega$, we let H_λ denote the irreducible $\mathfrak{k}$-module generated by the $\mathfrak{g} \times \mathfrak{g}'$-highest weight vector ϕ given in (7.20), (7.21) and (7.22). We also put $H_a = H_{\lambda_a}$.

9.2 THEOREM. Let $\lambda \in \Omega$ and $a = a(\lambda)$. The irreducible module $L(\lambda+\rho)$ in $\mathbb{H}_\lambda$ is precisely the kernel of the fundamental system $\mathcal{F}_a$.

PROOF. Let $F \in \mathbb{H}_\lambda$ and suppose $m(\partial_o)F = 0$ for all $m(\partial_o) \in \mathcal{F}_a$. Let $p \in P(\mathfrak{p}_+, \mathbb{C})$, $m(\partial_o) \in \mathcal{F}_a$, and $h \in H_{\lambda'}$, where λ' is given as in (8.3). By (2.8) we have $(F|pmh) = (m(\partial_o)F|ph) = 0$. We conclude $F \in L(\lambda+\rho)$ since J_λ is the orthogonal complement of $L(\lambda+\rho)$ and is spanned by the products pmh by (8.5).

Conversely, suppose $F \in L(\lambda+\rho)$. We proceed by considering cases.

SU(p,q) Since $\lambda \in \Lambda_a$, we have $Q(\lambda) = SU(p',q')$ where $1 \le p' \le p$ and $1 \le q' \le q$. Put $r = p-p'$, $s = q-q'$ and $k = \underline{k}(\lambda)$. Then r, s and k are the integers associated to $\lambda \in \Omega$ in (7.20). Let V be the space of functions defined in (7.9). We observe that H_a and H_λ are subspaces of V and that V is a K-invariant algebra. Since $V \subset H$ we know by (7.12) that IV is a $\mathfrak{g}$-module which decomposes into irreducible highest weight modules. Let P be orthogonal projection of $\mathbf{V}$ onto $\mathbb{H}_\lambda$ induced from orthogonal projection of V onto H_λ. Let $G \in \mathbf{V}$ and $m \in J_a^o$. Then mG is in $\mathbf{V}$ since V is closed under multiplication. Now $\Theta m = 0$ and by (7.6), Θ is multiplicative and so we have $\Theta mG = 0$. Thus $mG \in J(V)$ by (5.2) and by (5.1) $\mathsf{P}(mG) \in J_\lambda$. We now have $(m(\partial_o)F|G) = (F|mG) = (F|\mathsf{P}(mG)) = 0$ for all $G \in \mathbf{V}$. Since $m(\partial_o)F \in \mathbf{V}$, this implies $m(\partial_o)F = 0$.

Sp(n,ℝ) Since $\lambda \in \Lambda_a$ we have $Q(\lambda) = Sp(q,\mathbb{R})$ and $R(\lambda) = Sp(r,\mathbb{R})$ where $1 \le q \le r \le n$. Put $k = \underline{k}(\lambda)$ and let $\mathcal{A}_1$ and $\mathcal{A}_2$ be as in (7.23). Suppose $a \in \mathcal{A}_1$. Then k and $j = n - q$ are the integers associated to the $\mathfrak{g}\times\mathfrak{g}'$-weight vector ϕ given in (7.21)(1). Let V_1

be the space defined in (7.10) with $j = n-q$. Then H_a and H_λ are contained in V_1 and V_1 is a K-invariant algebra. Since $V_1 \subset H$ by (7.12) we know IV_1 is a $\mathfrak{g}$-module which decomposes into irreducible highest weight modules. Let $P: V_1 \to H_\lambda$ be the orthogonal projection induced from the projection of V_1 onto H_λ. Let $m(\partial_o) \in \mathcal{J}_a$ and $G \in V_1$. By (7.6) Θ is multiplicative and so by (5.2) we have $mG \in J(V_1)$ and by (5.1) we know $P(mG) \in J_\lambda$. Thus $(m(\partial_o)F|G) = (F|mG) = (F|P(mG)) = 0$ for all $G \in V_1$. This implies $m(\partial_o)F = 0$ since $m(\partial_o)F \in V_1$.

Suppose $a \in \mathcal{A}_2$. The integers $k = \underline{k}(\lambda)$ and $j = k-n+q$ are the integers associated to the weight vector in (7.21)(2). Let V_1 and V_2 be the spaces defined in (7.10) with $j = k-n+q$. Then H_λ and H_a are subspaces of V_2. Furthermore, V_2 is a K-invariant space of harmonics and, by definition of V_2, we have $V_1V_2 \subset V_2$. Let P be the orthogonal projection of V_2 onto H_λ. Let $m \in J_a^o$ and $G \in V_1$. Then $mG \in V_2$ and a straightforward calculation shows that $m(\partial_o)F \in V_1$. As above we have $mG \in J(V_2)$ and $(m(\partial_o)F|G) = (F|P(mG)) = 0$ by (5.1), for all $G \in V_1$. This implies $m(\partial_o)F = 0$.

SO*(2m) For $\lambda \in \Omega$ we have $Q(\lambda) = SO^*(2p)$, $3 \le p \le m$, or $Q(\lambda) = SU(1,1)$. Put $r = m-p$ if $Q(\lambda) = SO^*(2p)$ and $r = m-2$ if $Q(\lambda) = SU(1,1)$ and set $k = \underline{k}(\lambda)$. Then the integers r and k are the integers associated to λ in (7.22). Let V be the space defined in (7.11). The remainder of the proof proceeds as in the SU(p,q) case and is omitted.

This completes the proof of the theorem.

§10. Explicit Forms of the Systems of Differential Operators for the Classical Groups.

The canonical system $\mathcal{D}_\lambda$ is defined in terms of a basis of J_λ^o by replacing the variables in $\mathfrak{p}_+$ by differentiation with respect to those variables. However, the factorization theorem (6.15) shows that J_λ^o is determined in an elementary way by J_a^o for all λ in the cone Λ_a. Thus the K-module J_a^o plays a pivotal role in the construction of the canonical system $\mathcal{D}_\lambda$. Of course in section nine we found that the role of J_a^o is even more fundamental in the harmonic setting. For this harmonic setting the differential operators in the variables of both $\mathfrak{p}_+$ and M corresponding to J_a^o determine all the unitary representations for points in $\Lambda_a \cap \Omega$. So to complete our program we now describe bases for J_a^o for all the classical groups.

To be completely explicit we describe the results case by case in the subsections which follow. In most cases the results take the same form. We begin with a block matrix $x = \begin{pmatrix} A & B \\ C & 0 \end{pmatrix}$ where the entries of A are the variables of $\mathfrak{p}_+$ and the entries of B and C are taken from the variables of M. Then J_a^o has a basis given by determinants or Pfaffians of certain minors of the matrix x.

The remaining cases occur for $G = SO(2,n)$. Here the spin representation is needed to cover some rather special cases.

Unless otherwise specified, highest weights and highest weight vectors referred to below are with respect to the Borel and Cartan subalgebras defined in section seven.

Recall from (6.15) that μ_a denotes the highest weight of J_a^o. As in (6.19) we let H_λ^ℓ denote the space of homogeneous polynomials in $P(\mathfrak{p}_+, H_\lambda)$ of degree ℓ.

10.1 LEMMA. Let $a = (Q, R, \ell) \in \mathcal{A}$. Suppose J^o is a K-invariant subspace of H_a^ℓ which satisfies one of the following conditions:

A) $\lambda_a \in \Omega$ and J^o is contained in the kernel of Θ.

B) There exists a weight vector in J^o with highest weight μ_a and the dimension of J^o is less than or equal to the dimension of J_a^o.

Then $J^o = J_a^o$.

PROOF. Suppose J^o satisfies A. Then by (5.2) $J^o \subset J_a^o$. Since J_a^o is irreducible (cf. (3.1)) we obtain $J^o = J_a^o$. Suppose J^o satisfies B. By (6.8) a highest weight vector with weight μ_a is necessarily in J_a^o. Thus $J_a^o \subset J^o$ and dimension considerations imply $J_a^o = J^o$.

10.2 SU(p,q). Let $G = SU(p,q)$. Suppose $a = (Q, R, \ell) \in \mathcal{A}$. Then $Q = SU(p',q')$, $1 \le p' \le p$ and $1 \le q' \le q$. Let $r = p-p'$, $s = q-q'$, and $k = k_a = \ell - 1 + r + s$. By (6.22) we have $\lambda_a = \omega_{p-r} - (2+k)\omega_p + \omega_{p+s}$ and by (7.16) $\lambda_a \in \Lambda_r(k)$. By (7.20) the corresponding highest weight vector ϕ_a is $\phi_a(z) = \Delta_r(x)\varepsilon_s(y)$. In this case the K-span H_{λ_a} of ϕ_a has a particularly simple characterization: $f \in H_{\lambda_a}$ if an only if f is a sum of terms of the form $z \to \varphi_1(x)\varphi_2(y)$ where φ_1 is an r-minor of the first r columns of x and φ_2 is an s-minor of the last s columns of y.

10.2.1 Let $N(r,n) = \{i = (i_1,\ldots,i_r)\colon 1 \le i_1 < \ldots < i_r \le n\}$. Suppose $x \in M_{n,m}$. Then for $i \in N(r,n)$ and $j \in N(s,m)$ let x_{ij} denote the submatrix of x with rows and columns determined by i and j, respectively. We will let "$\bullet$" denote the full sequence $(1,\ldots,n)$. For $i = (i_1,\ldots,i_r) \in N(r,n)$ let $|i| = i_1 + \ldots + i_r$ and let i' be the sequence in $N(n-r,n)$ obtained by deleting $i_1,\ldots,i_r$ from $(1,\ldots,n)$. For any square matrix x let $|x|$ denote its determinant.

It will be convenient to record the following lemma concerning the expansion of determinants of a special form.

10.2.2 LEMMA. Let $x = \begin{pmatrix} u & v \\ w & 0 \end{pmatrix}$ where $u \in M_{n,m}$, $v \in M_{n,k}$, and $w \in M_{\ell,m}$. Assume $n > k$, $m > \ell$, and $n + \ell = m + k$. Then

$$|x| = \sum_{\substack{i \in N(n-k,n) \\ j \in N(m-\ell,m)}} (-1)^{|i|+|j|} |u_{ij}|\,|v_{i'\bullet}|\,|w_{\bullet j'}|$$

PROOF. This formula follows from the Laplace Expansion formula (cf. [p.93,38]).

Now fix $a \in \mathcal{A}$ and let p', q', r,s and k be as in (10.2). Let $A = A(z)$ be the first r columns of x and let $B = B(z)$ be the last s columns of y. Let $T \in \mathfrak{p}_+$. We define *a* ***minor of type-a*** to be the determinant of a $k+1 \times k+1$ submatrix of $\begin{pmatrix} T & A \\ B^t & 0 \end{pmatrix}$ with the understanding that the submatrix is chosen so that no column of A nor any row of B^t is omitted. More specifically, a minor of type-a is the determinant of a submatrix of the form

$$\begin{pmatrix} T_{ij} & A_{i\bullet} \\ (B^t)_{\bullet j} & 0 \end{pmatrix},$$

where $i \in N(k-s+1,p)$ and $j \in N(k-r+1,q)$. For a minor of type-a to exist it is necessary that the following inequalities hold: $k-r \geq 0$, $k-s \geq 0$, $k-r+1 \leq q$, and $k-s+1 \leq p$. The first two inequalities follow from the constraint $k \geq r+s$ which is implied by the formula for k above. The second two inequalities are equivalent to $\ell \leq \min(p',q')$ which follows from (6.3) since $\min(p',q')$ equals the split rank of Q.

10.2.3 PROPOSITION. The minors of type-a form a basis of J^o_a.

PROOF. Let J^o be the space spanned by the minors of type-a. It is easy to verify that the minors of type-a are linearly independent as functions on $\mathfrak{p}_+ \times M$. Let m be a minor of type-a. Applying (10.2.2) we see that $m \in \mathbb{H}^\ell_a$. Let J_1 be the $s \times k$ matrix $(0\ I_s)$ and J_2 the $k \times r$ matrix $\begin{pmatrix} I_r \\ 0 \end{pmatrix}$. Then $\begin{pmatrix} xy^t & A \\ B^t & 0 \end{pmatrix} = \begin{pmatrix} x \\ J_1 \end{pmatrix}(y^t\ J_2)$ has rank at most k since each factor in the product has rank at most k. Since m is a (k+1)-minor and $\theta(z) = xy^t$ it follows that $m(\theta(z),z) = 0$ for all $z \in M$. Thus $m \in \ker\Theta$ by (7.6). Let $(u,v) \in K$. The $\mathfrak{t}$-action of K is given by $\mathfrak{t}(u,v)F(T) = \omega(u,v)(F(u^{-1}Tv))$. It is straightforward to verify that J^o is K-invariant. From (10.1) we conclude that $J^o = J^o_a$.

10.3 $\mathbf{Sp(n,\mathbb{R})}$. Let $G = Sp(n,\mathbb{R})$. Suppose $a = (Q,R,\ell) \in \mathcal{A}$. Then $Q = Sp(q,\mathbb{R})$ and $R = Sp(r,\mathbb{R})$, $1 \leq q \leq r \leq n$. Let $k = k_a = 2n-(r+q-\ell+1)$. By (6.22) we have $\lambda_a = \omega_q + \omega_r - (2 + \frac{k}{2})\omega_n$ and by (7.16) $\lambda_a \in \Lambda_r(k)$. Recall the partition of $\mathcal{A}$ into two disjoint sets $\mathcal{A}_1$ and $\mathcal{A}_2$ and the corresponding formulas for ϕ_a given in (7.23). Recall also the notation of corollary (7.10). If $a \in \mathcal{A}_1$ let $A = A_1$ with $j = n-q$. If $a \in \mathcal{A}_2$ let $A = A_2$ with

$j = k-n+q$. In both cases A is an $n \times (n-q)$ matrix. Let B denote the first n-r columns of x. From (7.10) products of elements in V_1 with elements of V_2 are harmonic polynomials and so, we may apply (8.1) to conclude that H_{λ_a} is spanned by the products $f_1 f_2$ where f_1 is an (n-r)-minor of B and f_2 is an (n-q)-minor of A.

For $a \in \mathcal{A}$, a ***minor of type-a*** is defined to be the determinant of a $k+1 \times k+1$ submatrix of $\begin{pmatrix} T & A \\ B^t & 0 \end{pmatrix}$, $T \in \mathfrak{p}_+$, with the understanding that the submatrix is chosen so that no column of A nor any row of B^t is omitted. More specifically, a minor of type-a is the determinant of a submatrix of the form

$$\begin{pmatrix} T_{ij} & A_{i\bullet} \\ (B^t)_{\bullet j} & 0 \end{pmatrix},$$

where $i \in N(k+1-n+r,n)$ and $j \in N(k+1-n+q,n)$ (cf. (10.2.1) for notation). For a minor of type-a to exist it is necessary that the following inequalities hold: $k \geq n-q$, $k \geq n-r$, and $k+1 \leq \min(2n-r, 2n-q) = 2n-r$. The first two inequalities follow from the formula for $k = k_a$. The third inequality is equivalent to $\ell \leq q$ which follows from (6.3) since q equals the split rank of Q.

10.3.1 PROPOSITION. The linear span of the minors of type-a is J_a^o.

PROOF. Let J^o be the linear span of the minors of type-a and let m be a minor of type-a. It is immediate from (10.2.2) that m is an element of H_a^ℓ. We will now consider the case k is odd. The case k even is similar and the details will be left to the reader. Suppose $a \in \mathcal{A}_1$. Let J_1 be the $n-r \times d$ matrix $(I_{n-r} \;\; 0)$ and J_2 the $d \times (n-q)$ matrix $\begin{pmatrix} I_{n-q} \\ 0 \end{pmatrix}$. Let $z = (xyt)$ and recall that $\theta(z) = \frac{1}{2} zJz^t$. Then

$$\begin{pmatrix} zJz^t & A \\ B^t & 0 \end{pmatrix} = \begin{pmatrix} y & x & t \\ J_1 & 0 & 0 \end{pmatrix} \begin{pmatrix} x^t & 0 \\ y^t & J_2 \\ t^t & 0 \end{pmatrix}.$$

If $a \in \mathcal{A}_2$ let J_1 be as above but let J_2 be the $d \times (d-(k-n+q))$ matrix $\begin{pmatrix} 0 \\ I_{d-(k-n+q)} \end{pmatrix}$. Then

$$\begin{pmatrix} zJz^t & A \\ B^t & 0 \end{pmatrix} = \begin{pmatrix} y & x & t \\ J_1 & 0 & 0 \end{pmatrix} \begin{pmatrix} x^t & 0 & J_2 & 0 \\ y^t & I_d & 0 & 0 \\ t^t & 0 & 0 & 1 \end{pmatrix}.$$

In each case the rank is at most k since each factor on the right has rank at most k. Since m is a k+1 × k+1 minor of $\begin{pmatrix} T & A \\ B & 0 \end{pmatrix}$ it follows that $m(\theta(z),z) = 0$ for all $z \in M$. By (7.6) we know $m \in \ker \Theta$. Let $(u,t) \in K_2$ (cf. (7.5)). The $\mathfrak{t}$-action of K_2 is given by $\mathfrak{t}(u,t)F(T) = \omega(u,t)(F(u^{-1}T(u^t)^{-1}))$. It is then straightforward to show that J^o is K_2-invariant. It follows from (10.1) that $J^o = J^o_a$.

10.4 SO*(2m). Let $G = SO^*(2m)$. Suppose $a \in \mathcal{A}$ and suppose $\lambda_a \in \Omega$ is given as in (6.22). Then $Q = SO^*(2p)$, $3 \le p \le m$ or $Q = SU(1,1)$. Let $k = k_a$ be given as in (7.16) or (7.25). Then $k = \ell-1+m-p$, $1 \le \ell \le [p/2]$, $2 \le p \le m$; the case $p = 2$ corresponds to $Q = SU(1,1)$. Let $r = m-p$. By (7.22) $\phi_a(z) = \Delta_r(z)$ is the highest weight vector with highest weight λ_a. The space H_{λ_a} may be characterized as the space of r-minors in the first r columns of z.

Suppose x is a skew symmetric matrix of even order. Then the determinant of x as a polynomial in the matrix entries of x is a perfect square. The Pfaffian of x, Pf(x), is the polynomial in x whose square is the determinant of x and such that $\mathrm{Pf}\begin{pmatrix} 0 & I \\ -I & 0 \end{pmatrix} = 1$. The reader should consult Artin's book [4] for a list of the basic properties of Pfaffians.

Let A denote the first r columns of z and let $T \in \mathfrak{p}_+$. A ***Pfaffian of type-a*** is defined to be a Pfaffian of a $2(k+1) \times 2(k+1)$ skew symmetric submatrix of $\begin{pmatrix} T & A \\ -A^t & 0 \end{pmatrix}$ where it is understood that the submatrix is chosen so that no column of A (and hence no row of $-A^t$) is omitted. More specifically, a Pfaffian of type-a is the Pfaffian of a submatrix of the form

$$\begin{pmatrix} T_{ii} & A_{i\bullet} \\ -(A_{i\bullet})^t & 0 \end{pmatrix},$$

where $i \in N(2k-r+2,m)$ (cf. (10.2.1) for notation). For a Pfaffian of type-a to exist it is necessary that the following inequalities hold: $2k + 2 > r$ and $2(k+1) \le m + r$. The first inequality follows from the fact that $k \ge r$. The second inequality is equivalent to $2\ell \le p$ which follows from (6.3) since $[p/2]$ is the split rank of Q.

10.4.1 PROPOSITION. The Pfaffians of type-a form a basis of J_a^o.

PROOF. Let J^o be the space spanned by the Pfaffians of type-a. It is easy to verify that the Pfaffians of type-a are linearly independent. If we interchange two columns of $A_{i\bullet}$ and hence two rows of $A_{i\bullet}^t$ the Pfaffian of the block matrix above changes sign (cf. [4]). This implies that a Pfaffian of type-a is skew symmetric in the entries of A. Applying (10.2.1) we conclude that J^o is contained in $\mathbb{H}_a$. Let m be a Pfaffian of type-a. Then m is the Pfaffian of a matrix of the form $\begin{pmatrix} t & a \\ -a^t & 0 \end{pmatrix}$. Let $c \in \mathbb{C}$ with $c \neq 0$. Then

$$\begin{aligned} m^2(cT,z) &= \det \begin{pmatrix} ct & a \\ -a^t & 0 \end{pmatrix} \\ &= \det \begin{pmatrix} ct & cc^{-1}a \\ -a^t & 0 \end{pmatrix} = c^{2k-r+2} \det \begin{pmatrix} t & c^{-1}a \\ -a^t & 0 \end{pmatrix} \\ &= c^{2k-r+2}c^{-r} \det \begin{pmatrix} t & a \\ -a^t & 0 \end{pmatrix} = c^{2k-2r+2}m^2(T,z). \end{aligned}$$

It follows that m is homogeneous of degree $k - r + 1 = \ell$. Thus J^o is contained in $\mathbb{H}_a^\ell$. Let J be the $k \times r$ matrix given by $J = \begin{pmatrix} I_r \\ 0 \end{pmatrix}$. Then

$$\begin{pmatrix} xy^t - yx^t & A \\ -A^t & 0 \end{pmatrix} = \begin{pmatrix} x & y \\ 0 & J^t \end{pmatrix} \begin{pmatrix} y^t & J \\ -x^t & 0 \end{pmatrix}$$

has rank at most 2k since each matrix in the product has rank at most 2k. This implies $m(\theta(z),z) = 0$ for all $z \in M$. By (7.6) we know $m \in \ker\Theta$. Therefore J^o is contained in J_a^o. The formula for μ_a (cf. (6.22)) gives $\dim J_a^o = \binom{m}{p-2\ell} \leq \dim J^o$. It follows now that $J^o = J_a^o$.

10.4.2 For the remaining vertices we need to consider the double cover $\tilde{K}$ of K. Thus let $\tilde{K} = \{(u,t) \in K \times \mathbb{C}: t^2 = \det(u)\}$. The map $\rho: \tilde{K} \to K$ given by $\rho(u,t) = u$ is the covering map and the inverse image of Π, the Cartan subgroup of diagonal matrices in $K = U(m)$, is a Cartan subgroup of $\tilde{K}$, which we denote by $\tilde{\Pi}$.

10.4.3 The vertex $\omega_1 + \omega_{q+1} - (2m-q)\omega_m$, $2 \leq q \leq m-3$. Let $M = M_{m,m-1}$. Let H_1 be the space of m-1 minors of M and let H_2 be the space of m-q-1 minors of M which involve only the first m-q-1 columns. The theory of Young products (8.1) applies in this case and we find that $H_\lambda = H_1H_2$ is an irreducible $\tilde{K}$-module under the action τ given by $\tau(u,t)f(z) = t^{-(2m-q-4)}f(u^{-1}z)$. The highest weight of τ is λ_a. For $z \in M$ let $A = z$ and let $B = B(z)$ denote the first $m-q-1$ columns of z. Now let J^o be the space of functions spanned by the $(2m-q-1)$-minors of $\begin{pmatrix} T & A \\ B^t & 0 \end{pmatrix}$ where no column of A nor any row of B^t is omitted. More specifically, J^o is spanned by functions F_j, $j \in N(m-q,m)$, where F_j is the determinant of the submatrix

$$\begin{pmatrix} T_{\bullet j} & A \\ (B^t)_{\bullet j} & 0 \end{pmatrix}.$$

The functions F_j, $j \in N(m-q,m)$, are homogenous of degree 1 and have values in H_λ. The group $\tilde{K}$ acts on H_λ by $(t(u,t)F)(T) = \tau(u,t)F(u^{-1}T(u^{-1})^t)$. A straightforward calculation shows that J^o is $\tilde{K}$-invariant and has dimension $\le \binom{m}{m-q}$. However, if $j = (q+1, \ldots, m)$ then F_j is a highest weight vector with highest weight μ_a (cf. (6.22)). The $[\mathfrak{k},\mathfrak{k}]$-module with highest weight μ_a is the qth fundamental representation and so has dimension $\binom{m}{q}$. An application of (10.1)(B) gives $J^o = J^o_a$. Moreover, the set $\{F_j : j \in N(m-q,m)\}$ is a basis of J^o_a.

10.4.4 The vertex $\lambda_a = \omega_1 + \omega_{m-1} - (m+1)\omega_m$. Let $M = M_{m,m-1}$ and let H_1 be as in (10.4.3). Let H_2 be the space of linear functions of the first column of $z \in M$. Then $H_\lambda = H_1H_2$ is irreducible under the action τ of $\tilde{K}$ given by $\tau(u,t)f(z) = t^{-(m-2)} f(u^{-1}z)$. The highest weight of τ is λ_a. For $z \in M$, let $A = A(z) = z$ and let $B = B(z)$ denote the first column of z. Now let J^o be the space of functions spanned by the $(m+1)$-minors of $\begin{pmatrix} T & A \\ B^t & 0 \end{pmatrix}$ where no column of A is omitted. More specifically J^o is spanned by functions F_j, $j \in N(2,m)$, where F_j is the determinant of the submatrix

$$\begin{pmatrix} T_{\bullet j} & A \\ (B^t)_{\bullet j} & 0 \end{pmatrix}.$$

As in (10.4.3) $J^o = J_a^o$ and the set $\{F_j : j \in N(2,m)\}$ is a basis of J_a^o.

10.4.5 The vertex $\lambda_a = \omega_1 - (m-1)\omega_m$. Let $M = M_{m,m-1}$ and let H_a be the space of $(m-1)$-minors of M. Let τ denote the action of $\tilde{K}$ on H_a given by $\tau(u,t)f(z) = t^{-(m-3)}f(u^{-1}z)$. Then τ is irreducible with highest weight λ_a. Let J^o be the space of functions spanned by the m-minors of (T A) where no column of A is omitted. More specifically J^o is spanned by functions F_j, $j \in N(1,m)$, where F_j is the determinant of the submatrix $(T_{\bullet j}\ A)$. As in (10.4.3) $J^o = J_a^o$ and the set $\{F_j : j \in N(1,m)\}$ is a basis of J_a^o.

10.5 SO(2,2m-1). Let $n = 2m-1$ and let $G = SO(2,n)$ be the connected component of the identity of the group of real $(2+n) \times (2+n)$ matrices which preserve the form defined by the matrix $\begin{pmatrix} -I_2 & 0 \\ 0 & I_n \end{pmatrix}$. For $\theta \in \mathbb{R}$, let $t(\theta) = \begin{pmatrix} \cos\theta & -\sin\theta \\ \sin\theta & \cos\theta \end{pmatrix}$. The maximal compact subgroup $K = SO(2) \times SO(n)$ acts on $\mathfrak{p}^+ \cong \mathbb{C}^n$ by the formula $(t(\theta),D)\cdot T = e^{i\theta}TD^{-1}$. The complexification $so(n,\mathbb{C})$ of the Lie algebra of SO(n) is the space of complex skew symmetric matrices. We follow Knapp [32, pages 63, 74 and 75] in our choice of Cartan subalgebra and system of positive root vectors. Highest weights and highest weight vectors given below are computed with respect to these choices.

10.5.1 The vertex $\lambda_a = -(2m-p)\omega_1 + \omega_{p+1}$, $1 \le p \le m-2$. Let $M = M_{p,n}$ and let H^p be the space of p-minors of M. We define a representation τ of K on H^p by $\tau(t(\theta),D)f(z) = e^{-i\theta(2m-p-1)}f(zD)$. Then τ is irreducible and the highest weight is

λ_a. For $j \in N(p-1,n)$ (cf. (10.2.1) for notation) and $z \in M$ let z_j denote the $p\times(p-1)$ matrix consisting of the p-1 columns of z defined by j. Let J^o be the space of functions in $\mathbb{H}^p$ spanned by F_j, $j \in N(p-1,n)$, where $F_j(T,z) = \det(zT^t, z_j)$. Clearly J^o is K-invariant and $\dim J^o = \binom{n}{p-1}$. Now define p-1 tuples i and j by $i = (1,3,\ldots,2p-3)$ and $j = (2,4,\ldots,2p-2)$. A simple calculation shows that the function $F \in J^o$ defined by

$$F(T,z) = \det(zT^t, z_i - \sqrt{-1}\ z_j)$$

is a highest weight vector with weight $-(2m-p+1)\omega_1 + \omega_p = \mu_a$. In the case $p = 1$, J^o is simply the span of the function $(T,z) \mapsto zT^t$. An application of (10.1) gives $J^o = J^o_a$. Moreover, the set $\{F_j\colon j \in N(p-1,n)\}$ is a basis of J^o_a.

10.5.2 The vertex $\lambda_a = -(m+1)\omega_1 + 2\omega_m$. We retain the notation established in 10.5.1. We define a representation τ on H^{m-1} by $\tau(t(\theta),D)f(z) = e^{-im\theta}f(zD)$. Then τ is irreducible and has highest weight λ_a. We define J^o to be the span of the functions F_j, $j \in N(m-2,n)$. Then as in (10.5.1) we have $J^o = J^o_a$ and the set $\{F_j,\ j \in N(m-2,n)\}$ is a basis of J^o_a.

10.5.3 The vertex $\lambda_a = 0$. Let τ be the trivial representation on $\mathbb{C}$. Then J^o_a is the space of linear functions on $\mathfrak{p}_+$.

10.5.4 The vertex $\lambda_a = -(m-3/2)\omega_1$. We consider the two fold cover of SO(2) given by $\tilde{SO}(2) = \{(t(\theta),s)\colon \theta \in \mathbb{R},\ s^2 = e^{i\theta}\}$. Let $\tilde{K} = \tilde{SO}(2)\times SO(n)$. Let τ be the one dimensional representation of $\tilde{K}$ on $\mathbb{C}$ given by $\tau((t(\theta),s),D) = s^{-(2m-3)} = e^{-i(2m-3)\theta/2}$. Then the highest weight of τ is λ_a. A simple argument shows that the function $T \rightarrow T\cdot T = (T_1)^2 + \ldots + (T_n)^2$ has highest weight μ_a.

Applying (10.1) we conclude J_a^o is the span of the function $T \to T \cdot T$.

10.5.5 The half-spin representations. Here we make some general remarks about the half spin representations. Let n be any positive integer and let $\mathbb{C}_n$ be the complexification of the real Clifford algebra associated with the quadratic form $Q(x) = -|x|^2$ on $\mathbb{R}^n$. Let $e_1, \ldots, e_n$ be the standard orthonormal basis of $\mathbb{R}^n$. For $s = \{s_1, \ldots, s_k\} \subset \{1,2, \ldots, n\}$, $s_1 \leq \ldots \leq s_k$, we denote $e_{s_1} \cdots e_{s_k}$ by e_s and we put $e_o = 1$. The collection of all such vectors forms a basis of $\mathbb{C}_n$. Let $\mathbb{C}_n^o$ be the span of those e_s, $s=(s_1, \ldots, s_k)$ with k even. Then $\mathbb{C}_n^o$ forms a subalgebra isomorphic to $\mathbb{C}_{n-1}$. We regard $\mathbb{R}^n$ as a subset of $\mathbb{C}_n$ and let S^{n-1} denote the unit sphere in $\mathbb{R}^n$. Let $j = [n/2]$. The set of elements in $\mathbb{C}_n^o$ of the form $a_1 \cdots a_{2k}$, where $a_i \in S^{n-1}$ and $1 \leq k \leq j$, is the group Spin(n). The map ρ: Spin(n)$\to$SO(n) given by $\rho(a)x = axa^{-1}$, $x \in \mathbb{R}^n$, is the covering map.

Let $I = (e_1 - ie_2) \cdots (e_{2j-1} - ie_{2j})$. Then $e_{2k}I = - ie_{2k-1}I$, $1 \leq k \leq j$. Suppose $\mathcal{B}_e$ (resp. $\mathcal{B}_o$) $= \{e_sI$: card(s) is even (resp. odd) and $s_i \in S$ is an odd integer$\}$. Let S_e and S_o be the span of B_e and $\mathcal{B}_o$ respectively. Clearly S_e and S_o are invariant under the left action of $\mathbb{C}_n^o$. Moreover, if $x \in \mathbb{R}^n$, $x \neq 0$, then $xS_e = S_o$ and $xS_o = S_e$. Clearly, the dimensions of S_e and S_o are less than or equal to 2^{j-1} (resp. 2^j) if n is even (resp. n is odd). By Propositions 6.14, p. 288 and 6.18, p. 289 of [7] S_e and S_o are the half-spin representations of Spin(n) which are equivalent if n is odd and inequivalent if n is even.

10.5.6 The vertex $\lambda_a = -(m-1/2)\omega_1 + \omega_m$. We retain the notation of the previous section. Let $n = 2m-1$ and let $\tilde{K} = SO(2)$

$\times$ spin(n). We define a representation τ of $\tilde{K}$ on S_e by the formula $\tau(t(\theta),a)x = e^{-i\theta(m-1)}ax$, $a \in$ spin(n), $x \in S_e$. It is readily verified that τ has highest weight λ_a. Let J^o denote the vector space of functions F_y, $y \in S_o$, defined by $F_y(T) = Ty = \Sigma\, T_i\, e_i y$. Then F_y is an S_e-valued polynomial of degree one on $\mathfrak{p}_+$. If $(t(\theta),a) \in \tilde{K}$ then $t(t(\theta),a)F_y = e^{-im\theta}F_{ay}$. Thus it is easy to see that J^o is an irreducible $\tilde{K}$-space and if y_0 is the highest weight vector of S_o then F_{y_0} is the highest weight vector for J^o. Since S_o is equivalent to S_e, J^o has highest weight μ_a. It follows from (10.1) that $J^o = J^o_a$.

10.6 SO(2,2m-2). Let $G = SO(2,n)$, where $n = 2m-2$. The comments in the introductory paragraph of (10.5) apply when n is even. We thus retain the notation established there.

10.6.1 The vertex $\lambda_a = -(2m-p-1)\omega_1 + \omega_{p+1}$, $1 \le p \le m-3$. Let M and H^p be as in (10.5.1). We define a representation τ of K on H^p by $\tau(t(\theta),D)f(z) = e^{-i\theta(2m-p-2)}\, f(zD)$. Then τ is irreducible and the highest weight is λ_a. As in (10.5.1) J^o_a is the space of functions spanned by the basis $\{F_j\colon j \in N(p-1,n)\}$.

10.6.2 The vertex $\lambda_a = -(m+1)\omega_1 + \omega_{m-1} + \omega_m$. We define a representation τ of K on H^{m-2} by $\tau(t(\theta),D)f(z) = e^{-im\theta}$ $f(zD)$. As in (10.5.2) J^o_a is the space of functions spanned by the basis $\{F_j\colon j \in N(m-2,n)\}$.

10.6.3 The vertex $\lambda_a = -(m-1)\omega_1 + \omega_m$. We retain the notation of (10.5.5). Let $\tilde{K} = \tilde{SO}(2) \times$ Spin(n), where $\tilde{SO}(2)$ is the two-fold cover of SO(2) as given in (10.5.4). We define a representation τ of $\tilde{K}$ on S_e by $\tau((t(\theta),s),a)x = s^{-2m+3}$ ax. Then τ is irreducible and has highest weight λ_a. Let J^o be the space of functions F_x, $x \in S_o$, defined by $F_x(T) = Tx$. Clearly, J^o consists of homogeneous polynomials of degree 1 with values in S_e. If $((t(\theta),s),a) \in \tilde{K}$ then $t((t(\theta),s),a)F_x = s^{-2m+1} F_{ax}$. Thus J^o is an irreducible $\tilde{K}$-space and if x is the highest weight vector of S_o then F_x is the highest weight vector for J^o with highest weight μ_a. It follows from (10.1) that $J^o = J^o_a$.

10.6.4 The vertex $\lambda_a = 0$. As in (10.5.3) J^o_a is the space of linear functions on $\mathfrak{p}_+$.

10.6.5 The vertex $\lambda_a = -(m-2)\omega_1$. As in (10.5.4) J^o_a is the span of the function $T \to T \cdot T$.

10.6.6 The vertex $\lambda_a = -(m-1)\omega_1 + \omega_{m-1}$. Reversing the roles of S_e and S_o in (10.6.3) produces J^o_a for this case.

§11. The Ladder Representation Examples

We now describe the fundamental systems associated to the irreducible components of the oscillator representation of $SU(p,q)$, $Sp(n,\mathbb{R})$ and $SO^*(2m)$ when the integer k from section seven equals one. These components are called ladder representations and are a well-known series of representations in physics (cf. for example [29], [36] and [37]). We obtain from the previous section three (resp. two) fundamental systems for $SU(p,q)$ (resp. $Sp(n,\mathbb{R})$). For the $SO^*(2m)$ case, there are two fundamental systems when $m \geq 4$ and one such system for $m = 3$. These eight systems are then sufficient to determine all of the ladder representations.

From (7.18) and (7.25) we calculate the possible a with $k_a = 1$. The resulting eight cases are labeled a_i, $1 \leq i \leq 8$, and described by the following table.

TABLE 11.1

Case	Group	Highest Weights	d	Q	R	ℓ
a_1	$SU(p,q)$	$d\omega_{p-1} - (1+d)\omega_p$	$d\geq1$	$SU(p-1,q)$	$SU(p-1,q)$	1
a_2		$-(1+d)\omega_p + d\omega_{p+1}$	$d\geq1$	$SU(p,q-1)$	$SU(p,q-1)$	1
a_3		$-\omega_p$		$SU(p,q)$	$SU(p,q)$	2
a_4	$Sp(n,\mathbb{R})$	$\omega_{n-1} - \frac{3}{2}\omega_n$		$Sp(n-1,\mathbb{R})$	$Sp(n,\mathbb{R})$	1
a_5		$-\frac{1}{2}\omega_n$		$Sp(n,\mathbb{R})$	$Sp(n,\mathbb{R})$	2
a_6	$SO^*(2m)$	$d\omega_{m-1} - (2+d)\omega_m$	$d\geq1$	$SO^*(2m-2)$	$SO^*(2m-2)$	1
a_7	$m \geq 4$	$-2\omega_m$		$SO^*(2m)$	$SO^*(2m)$	2
a_8	$SO^*(6)$	$d\omega_2 - (2+d)\omega_3$	$d\geq1$	$SU(1,1)$	$SU(1,1)$	1

In each of these eight cases we define $\mathfrak{k}$-submodules H_i, $1 \leq i \leq 8$, of the space of harmonic polynomials as follows. Let H_i for $i = 3$, 5 and 7 equal $\mathbb{C}$, the algebra of constant functions on M. Let H_1, H_6 and H_8 denote the sum of the homogeneous polynomials on M of positive degree in the variables of the coordinates of the column vector x. Let H_2 be the corresponding $\mathfrak{k}$-submodule in the coordinates of the column vector y. Lastly set H_4 equal to the degree one homogeneous polynomials in the coordinates of z ($z = t$ when $k = 1$).

Next, for each of these cases we describe a set $\mathcal{F}_i$ of differential operators in the variables of $\mathfrak{p}_+ \times M$. We keep our standard notation as in section seven, i.e., the matrix entries T_{ij} are the coordinates of $\mathfrak{p}_+$ and the matrix entries of z (or its expressions in terms of x, y and t) are the variables of M. Here for $k = 1$, x,y and t are column vectors so we shall label the matrix entries by the row index alone. For any matrix (A_{ij}), let $(A_{ij})(\partial_0)$ denote the matrix $(\frac{\partial}{\partial A_{ij}})$ of the same size. With this notation we can define the sets $\mathcal{F}_i$ as follows:

Table 11.2

System	Operators
$\mathcal{F}_1$:	$\left(\dfrac{\partial}{\partial x_j}\ \dfrac{\partial}{\partial T_{ir}} - \dfrac{\partial}{\partial x_i}\ \dfrac{\partial}{\partial T_{jr}}\right)$, $1 \le i < j \le p$, $1 \le r \le q$
$\mathcal{F}_2$:	$\left(\dfrac{\partial}{\partial y_j}\ \dfrac{\partial}{\partial T_{ri}} - \dfrac{\partial}{\partial y_i}\ \dfrac{\partial}{\partial T_{rj}}\right)$, $1 \le i < j \le q$, $1 \le r \le p$
$\mathcal{F}_3$:	all 2×2 minors of $(T_{ij})(\partial_0)$
$\mathcal{F}_4$:	$\left(\dfrac{\partial}{\partial t_j}\ \dfrac{\partial}{\partial T_{ir}} - \dfrac{\partial}{\partial t_i}\ \dfrac{\partial}{\partial T_{jr}}\right)$, $1 \le i < j \le n$, $1 \le r \le n$
$\mathcal{F}_5$:	all 2×2 minors of $(T_{ij})(\partial_0)$
$\mathcal{F}_6$:	$\left(\dfrac{\partial}{\partial x_i}\ \dfrac{\partial}{\partial T_{jr}} + \dfrac{\partial}{\partial x_j}\ \dfrac{\partial}{\partial T_{ri}} + \dfrac{\partial}{\partial x_r}\ \dfrac{\partial}{\partial T_{ij}}\right)$, $1 \le i < j < r \le m$
$\mathcal{F}_7$:	$\left(\dfrac{\partial}{\partial T_{it}}\ \dfrac{\partial}{\partial T_{js}} + \dfrac{\partial}{\partial T_{jt}}\ \dfrac{\partial}{\partial T_{si}} + \dfrac{\partial}{\partial T_{st}}\ \dfrac{\partial}{\partial T_{ij}}\right)$, $1 \le i < j < s < t \le m$
$\mathcal{F}_8$:	$\left(\dfrac{\partial}{\partial x_1}\ \dfrac{\partial}{\partial T_{23}} + \dfrac{\partial}{\partial x_2}\ \dfrac{\partial}{\partial T_{31}} + \dfrac{\partial}{\partial x_3}\ \dfrac{\partial}{\partial T_{12}}\right)$

11.3 PROPOSITION. As usual let $\mathbb{H}_i = P(\mathfrak{p}_+, H_i)$, $1 \le i \le 8$. Then the kernel of the system $\mathcal{F}_i$ in $\mathbb{H}_i$ is the (multiplicity free) sum of all the ladder representations with highest weights in the cone

Λ_{a_i}, $1 \le i \le 8$. Moreover, the individual simple modules in H_i correspond to the decomposition of H_i by homogeneous degree.

This is an immediate consequence of (9.2) and the description of the fundamental systems in section ten.

11.4 REMARKS. 1) The case a_8 is of course related to the group SU(1,3) since this group and SO*(6) are locally isomorphic. A calculation shows that the unitary representations occurring in the a_8 case occur naturally in the harmonic setting for SU(1,3) when $k = 2$. So here we are obtaining additional ladder representations for SU(1,3) from the local isomorphism with SO*(6).

2) L. Mantini considers cohomological constructions of the ladder representations for U(p,q) in [36] and [37]. She gives systems of differential equations, each of which is of the form $m(\partial)f = 0$, $m \in J$. Her systems properly contain those given in $\mathcal{F}_1$ and $\mathcal{F}_2$ in Table (11.2).

3) A group invariant differential operator D_λ, $\lambda \in \Lambda$, was studied in [10]. It was shown that for ladder representations of SU(p,q), the kernel of D_λ satisfied the equations given in (11.2). Thus the results of this article imply that the space of K-finite vectors in $\mathrm{Ker}(D_\lambda)$ is an irreducible $\mathfrak{g}$-module.

§12. $K_{\mathbb{C}}$-Orbits in $\mathfrak{p}_+$ and the Wallach Representations

In this section we consider the natural association between $K_{\mathbb{C}}$-orbits in $\mathfrak{p}_+$ and a class of unitarizable highest weight representations. This association was described some years ago in the article by Harris and Jakobsen [19] (cf. also [44]). Thus the results given below are not new. They follow, however, as a simple application of the general theory we have developed and so we include them as an illustrative set of examples. To briefly summarize, recall the formula for θ given in the harmonic setting in the paragraph after (7.2). If R is the closure of any $K_{\mathbb{C}}$-orbit in $\mathfrak{p}_+$ then R is the image of θ for some k. This fact then implies that the coordinate ring of R supports a $\mathfrak{g}$-structure which is irreducible and unitarizable. We identify the resulting representations as those studied by Wallach [45] and [46].

Let ζ be defined as in section six. Then $\zeta = \omega_o$ is the fundamental dominant weight corresponding to the noncompact simple root. Recall the definition of Λ and the line $L(\omega_o)$ as given in section six. The set $L(\omega_o) \cap \Lambda$ is referred to in the literature as the *Wallach set* and corresponds to the scalar valued representations extensively studied by Wallach in [45] and [46]. We are particularly interested in the reduction points of the Wallach set. Put $\Lambda_W = L(\omega_o) \cap \Lambda_r$. Let $\lambda \in \Lambda_W$ and $a = a(\lambda)$. We then have $a = (\Delta, \Delta, \ell)$, $1 \le \ell \le$ split rank of G. Furthermore, λ is the vertex $\lambda_a = -C(\ell-1)\zeta$, where C is given in Table 6.20.

Suppose G equals SU(p,q), Sp(n,$\mathbb{R}$) or SO*(2m). If $k = k_a$ (cf. (7.25)) then $k = \ell-1$. We identify H_λ with the complex numbers so that $\mathbb{H}_\lambda = P(\mathfrak{p}_+)$. We let $P = P(\mathfrak{p}_+)$, $L = L(\lambda+\rho)$, and $J = J_\lambda$. The $\mathfrak{g}$-subspace in F with highest weight λ is the algebra of invariant polynomials I. The restriction of the map Θ to P, which we will also denote by Θ, is a surjection onto I and is given by $\Theta p = p\circ\theta$, (cf. (7.6)). The kernel of Θ is J and since J is a module over $P(\mathfrak{p}_+)$, it is therefore an ideal.

Let R_k denote the set of rank k (resp. 2k) or less matrices of $\mathfrak{p}_+$ for G = SU(p,q) or Sp(n,$\mathbb{R}$) (resp. SO*(2m)). Note here that if G = SU(p,q) then R_k is a determinantal variety in $M_{p,q}$ and if G = Sp(n,$\mathbb{R}$) or SO*(2m) then R_k is a linear section of a determinantal variety in $M_{n,n}$ or $M_{m,m}$, respectively (cf. [3]).

12.1. PROPOSITION. The set R_k is the image of the map $\theta: M \to \mathfrak{p}_+$ and is the closure of the $K_{\mathbb{C}}$-orbit consisting of rank k (resp. 2k) matrices.

PROOF. The proof that $R_k = \theta(M)$ for the SU(p,q) case (resp. Sp(n,$\mathbb{R}$), SO*(2m) cases) follows from the singular value decomposition of $M_{p,q}$ (resp. from the polar decompositions [p.63,67;25] of $\mathfrak{p}_+$).

12.2 PROPOSITION. The ideal of R_k is J and is generated by the k+1-minors (resp. 2(k+1)-Pfaffians) of $\mathfrak{p}_+$ for G = SU(p,q) or Sp(n,$\mathbb{R}$) (resp. SO*(2m)).

PROOF. This follows from the description of J given in (10.2) to (10.4).

12.3 REMARK. If $G = SU(p,q)$ the above theorem is known as the second fundamental theorem of invariant theory (cf. p.71;[3]).

For completeness, we include a representation theoretic proof of the following known result.

12.4 PROPOSITION. The algebraic set R_k is an irreducible variety.

PROOF. Suppose f and g are in P. We write $f = f_1 + f_2$ and $g = g_1 + g_2$ where $f_1, g_1 \in L$ and $f_2, g_2 \in J$. Then $fg \in J$ if and only if $f_1g_1 \in J$. Suppose then that $f_1g_1 \in J$ and $g_1 \neq 0$. The restriction of Θ to L is an injective map since J, the orthogonal complement of L, is the kernel of Θ. Furthermore, $0 = \Theta(f_1g_1) = \Theta(f_1)\ \Theta(g_1)$. Since $g_1 \neq 0$, $\Theta(g_1) \neq 0$. This implies $\Theta f_1 = 0$ and hence $f_1 = 0$. But this implies $f \in J$. Therefore R_k is an irreducible variety.

12.5 PROPOSITION. The coordinate ring $C_k = P/J$ of R_k has a $\mathfrak{g}$-structure which makes C_k an irreducible unitarizable $\mathfrak{g}$-module isomorphic to the algebra I of invariants (for the dual group) in Fock space.

PROOF. Let $C_k = P/J$ be the coordinate ring of R_k. The $\mathfrak{k}^*$-action of $\mathfrak{g}$ on P and J induces a $\mathfrak{g}$-action on C_k. The map $\tilde{\Theta}$: $C_k \to I$ given by $\tilde{\Theta}(f + J) = \Theta f$ is clearly a well defined $\mathfrak{g}$-module isomorphism.

12.6 REMARK. Let $G = SU(p,q)$. When regarded as vector spaces the isomorphism $C_k \cong I$ is known as the first fundamental theorem of invariant theory (cf. p.77 [3]).

Notational Index

Sec	Notation / Symbols
1	$\mathfrak{g}_o$, $\mathfrak{k}_o$, $\mathfrak{h}_o$, $\mathfrak{g}$, $\mathfrak{k}$, $\mathfrak{h}$, $\mathfrak{p}_-$, $\mathfrak{p}_+$, $\mathfrak{h}^*$, θ, Δ, Δ_c, Δ_n, Δ^+, ρ, $\leq$, $\mathcal{L}^+$, $\mathcal{W}$, $\mathcal{W}_c$, s_α, $G_{\mathbb{C}}$, $K_{\mathbb{C}}$, P_+, P_-, $\mathfrak{E}$, Γ, t, t_1, t_2, $\mathfrak{O}_E$, $P(V,W)$, $S(V)$, $P(\mathfrak{p}_+,E)$, E, $\mathfrak{q}$, $\bar{\mathfrak{q}}$, σ
2	$d\mu$, $G(\partial)$, L, J
3	$H^i(\mathfrak{p}_+,A)$, $F(\xi)$, $\mathcal{D}_L$, $\mathcal{D}_\lambda$, $\mathcal{O}$, $\mathcal{O}(\mathfrak{g},\mathfrak{q})$, $\mathcal{W}_\lambda$, Δ_λ, $\Delta_{\lambda,c}$, Δ^+_λ, $\Delta^+_{\lambda,c}$, ℓ_λ, $\mathcal{W}_{\lambda,c}$, $\mathcal{W}^{c,i}_\lambda$, $(\mathfrak{g}_\lambda,\mathfrak{k}_\lambda)$, ξ, $\mathbb{N}^*$
4	$\mathcal{O}(\lambda)$
5	$\mathbb{H}$, $J(H)$, P_V, F, $\mathbb{F}$, V
6	$\mathfrak{h}^*_{\mathbb{R}}$, Λ, Λ_r, $L(\lambda)$, $\ell(\lambda)$, $Q(\lambda)$, $R(\lambda)$, J_λ, $a(\lambda)$, Λ_a, $\mathcal{A}$, C_a, λ_a, P, H_a, H', H_λ, μ_a, $\overset{\circ}{J}_a$, $\overset{\circ}{J}_\lambda$, $\mathbb{H}^\ell_\lambda$, ω_j
7	ω, (G,G'), $M_{r,s}$, M, F, $\mathbb{F}$, θ, Θ, I, H, $\lambda(.)$, H^σ, $\Sigma(k)$, $\Lambda(k)$, $\Lambda_r(k)$, $\underline{k}(\lambda)$, Ω, k_a, $\mathfrak{b}_k$, $\mathfrak{h}'$, $\mathfrak{b}'$, $\mathfrak{b}(n)$, ε_j, Δ_i, $\tilde{\Delta}_j$, $\mathcal{A}_1$, $\mathcal{A}_2$
8	H_λ, H', H_a
9	$\mathcal{F}_a$, $\mathcal{D}_a$, ∂_o
10	$N(r,n)$, $Pf(x)$, C_n, $\overset{\circ}{C}_n$, S_e, S_o
11	
12	Λ_W, R_k, C_k

Bibliography

1 J. Adams, Unitary highest weight modules, *Advances in Math.* 63 (1987), 113-137.

2 F. W. Anderson and K. R. Fuller, *Rings of Categories of Modules*, Springer-Verlag, New York, 1974.

3 E. Arbarello, M. Cornalba, P. A. Griffiths and J. Harris, *Geometry of Algebraic Curves*, Vol. 1, Springer-Verlag, New York, 1985.

4 E. Artin, *Geometric Algebra*, Interscience Publishers Inc., New York, 1957.

5 D. Barbasch, Filtrations on Verma modules, *Ann. Sci. Ecol. Norm. Sup.* 16 (1983), 489-494.

6 B. Boe and D. Collingwood, A multiplicity one theorem for holomorphically induced representations, *Math. Zeit.* 192 (1986), 252-282.

7 T. Brocker and T. tom Dieck, *Representations of Compact Lie Groups*, Springer-Verlag, New York, 1985.

8 D. Collingwood, The n-homology of Harish-Chandra modules: generalizing a theorem of Kostant, *Math. Ann.* 272 (1985), 161-187.

9 M. G. Davidson, The harmonic representation of U(p,q) and its connection with the generalized unit disk, *Pacific J. Math.* 129 (1987), 33-55.

10 M. G. Davidson and R. J.Stanke, Gradient-type differential operators and unitary highest weight representations of SU(p,q), *J. Func. Anal.* 81 (1988), 100-125.

11 J. Dixmier, *Algebres Enveloppantes*, Gauthier-Villars, Paris, 1974.

12 T. J. Enright, Analogues of Kostant's u-cohomology formulas for unitary highest weight modules, *J. Reine Angew. Math.* 392 (1988), 27-36.

13 T. J. Enright and A. Joseph, An intrinsic analysis of unitarizable highest weight modules, to appear in Math. Ann.

14 T. J. Enright and R. Parthasarathy, A proof of a conjecture of Kashiwara and Vergne, Springer-Verlag Lecture Notes in Math. 880 (1981).

15 T. J. Enright, R. Howe and N. Wallach, A classification of unitary highest weight modules in "Proceedings, Representation Theory of Reductive Groups (Park City, UT, 1982), pp. 97-149; Progr. Math. 40(1983) 97-143.

16 T. J. Enright and B. Shelton, Categories of highest weight modules: applications to classical Hermitian symmetric pairs, *Memoirs Amer. Math. Soc.* 367 (1987).

17 T. J. Enright and B. Shelton, Highest weight modules for Hermitian symmetric pairs of exceptional type, *Proc. Amer. Math. Soc.* 105 (1989), 1-13.

18 Harish-Chandra, Representations of semisimple Lie groups, IV, V and VI, *Amer. J. Math.* 77 and 78 (1955) and (1956), 743-777, 1-41 and 564-628.

19 M. Harris and H. P. Jakobsen, Singular holomorphic representations and singular modular forms, *Math. Ann.* 259 (1982), 227-244.

20 M. Harris and H. P. Jakobsen, Covariant differential operators, Springer-Verlag Lecture Notes in Physics 180, 1983.

21 H. Hecht, The characters of some representations of Harish-Chandra, *Math. Ann.* 219 (1976), 213-226.

22 R. Howe, Dual pairs in physics: harmonic oscillators, photons, electrons and singletons, Lectures in Applied Mathematics Vol. 21, Amer. Math. Soc., Providence, R.I., 1985.

23 R. Howe, Remarks on classical invariant theory, *Trans. Amer. Math. Soc.* 313 (1989), 539-570.

24 R. Howe, Transending the classical invariant theory, *J. Amer. Math Soc.* 2 (1989), 535-552.

25 L. K. Hua, *Harmonic Analysis of Functions of Several Complex Variables in the Classical Domains*, Amer. Math. Soc., Providence, R.I., 1963; revised 1979.

26 H. P. Jakobsen, The last possible place of unitarity for certain highest weight modules, *Math. Ann.* 256 (1981), 439-447.

27 H. P. Jakobsen, Hermitian symmetric spaces and their unitary highest weight modules, *J. Func. Anal.* 52 (1983), 385-412.

28 H. P. Jakobsen, Basic covariant differential operators on Hermitian symmetric spaces, *Ann. Sci. Ecol. Norm. Sup.* 18 (1985), 412-436.

29 H. P. Jakobsen and M. Vergne, Restrictions and expansions of holomorphic representations, *J. Func. Anal.* 34 (1979), 29-53.

30 J. C. Jantzen, *Moduln mit einem höchsten Gewicht*, Springer-Verlag Lecture Notes in Math. 750 (1979).

31 M. Kashiwara and M. Vergne, On the Segal-Shale-Weil representations and harmonic polynomials, *Invent. Math.* 44 (1978), 1-47.

32 A. W. Knapp, *Representation Theory of Semisimple Lie Groups, An Overview Based on Examples*, Princeton University Press, Princeton, N.J., 1986.

33 B. Kostant, Lie group representations on rings of polynomials, *Amer. J. Math.* 85 (1963), 327-404.

34 B. Kostant, Verma modules and the existence of quasi-invariant differential operators, Springer-Verlag Lecture Notes in Math. 466 (1975).

35 G. Mack and I. Todorov, Irreducibility of the ladder representations of U(2,2) when restricted to the Poincare subgroup, *J. Math. Phys.* 10 (1969), 2078-2085.

36 L. A. Mantini, An integral transform in L^2-cohomology for the ladder representations of U(p,q), *J. Func. Anal.* 60 (1985), 211-242.

37 L. A. Mantini, An L^2-cohomology construction of negative spin mass zero equations for U(p,q), *J. Math. Anal. and Appl.* 136 (1988), 419-449.

38 T. Muir, *A Treatise on the Theory of Determinants*, Dover Publications Inc., New York, 1960.

39 M. A. Naimark and A. I. Stern, *Theory of Group Representations*, Springer-Verlag, New York, 1982.

40 K. R. Parthasarathy, R. Ranga Rao and V. S. Varadarajan, Representations of complex semi-simple Lie groups and Lie algebras, *Ann. Math.* 85 (1967), 383-429.

41 A. Rocha-Caridi and N. R. Wallach, Projective module over Graded Lie Algebras, I, *Math. Zeit.* 180(1982), 151-177.

42 C. M. Patton and H. Rossi, Unitary structures on cohomology, *Trans. Amer. Math. Soc.* 290 (1985), 235-257.

43 W. Schmid, Die randwerte holomorpher funktionen auf hermitesch symmetrischen raumen, *Invent. Math.* 9 (1969/70), 61-80.

44 M. Vergne and H. Rossi, Analytic continuation of the holomorphic discrete series of a semi-simple Lie group, *Acta Math.* 136 (1976), 1-59.

45 N. Wallach, The analytic continuation of the discrete series I, *Trans. Amer. Math. Soc.* 251 (1979), 1-17.

46 N. Wallach, The analytic continuation of the discrete series II, *Trans. Amer. Math. Soc.* 251 (1979), 19-37.

47 D. P. Zelobenko, *Compact Lie Groups and Their Representations*, Amer. Math. Soc., Providence, R.I., 1973.

Department of Mathematics
Louisiana State University
Baton Rouge, Louisiana 70803

Department of Mathematics
University of California, San Diego
La Jolla, California 92093

Department of Mathematics
Baylor University
Waco, Texas 76798

MEMOIRS of the American Mathematical Society

SUBMISSION. This journal is designed particularly for long research papers (and groups of cognate papers) in pure and applied mathematics. The papers, in general, are longer than those in the TRANSACTIONS of the American Mathematical Society, with which it shares an editorial committee. Mathematical papers intended for publication in the Memoirs should be addressed to one of the editors:

General instructions to authors for

PREPARING REPRODUCTION COPY FOR MEMOIRS

> **For more detailed instructions send for AMS booklet, "A Guide for Authors of Memoirs." Write to Editorial Offices, American Mathematical Society, P.O. Box 6248, Providence, R.I. 02940.**

MEMOIRS are printed by photo-offset from camera copy fully prepared by the author. This means that the finished book will look exactly like the copy submitted. Thus the author will want to use a good quality typewriter with a new, medium-inked black ribbon, and submit clean copy on the appropriate model paper.

Model Paper, provided at no cost by the AMS, is paper marked with blue lines that confine the copy to the appropriate size.

Special Characters may be filled in carefully freehand, using dense black ink, or **INSTANT** ("rub-on") **LETTERING** may be used. These may be available at a local art supply store.

Diagrams may be drawn in black ink either directly on the model sheet, or on a separate sheet and pasted with rubber cement into spaces left for them in the text. Ballpoint pen is not acceptable.

Page Headings (Running Heads) should be centered, in CAPITAL LETTERS (preferably), at the top of the page — just above the blue line and touching it.

LEFT-hand, EVEN-numbered pages should be headed with the AUTHOR'S NAME;

RIGHT-hand, ODD-numbered pages should be headed with the TITLE of the paper (in shortened form if necessary).

Exceptions: PAGE 1 and any other page that carries a display title require NO RUNNING HEADS.

Page Numbers should be at the top of the page, on the same line with the running heads.

LEFT-hand, EVEN numbers — flush with left margin;

RIGHT-hand, ODD numbers — flush with right margin.

Exceptions: PAGE 1 and any other page that carries a display title should have page number, centered below the text, on blue line provided.

FRONT MATTER PAGES should be numbered with Roman numerals (lower case), positioned below text in same manner as described above.

MEMOIRS FORMAT

> **It is suggested that the material be arranged in pages as indicated below.**
> **Note: Starred items (*) are requirements of publication.**

Front Matter (first pages in book, preceding main body of text).

Page i — *Title, *Author's name.

Page iii — Table of contents.

Page iv — *Abstract (at least 1 sentence and at most 300 words).

Key words and phrases, if desired. (A list which covers the content of the paper adequately enough to be useful for an information retrieval system.)

**1991 Mathematics Subject Classification*. This classification represents the primary and secondary subjects of the paper, and the scheme can be found in Annual Subject Indexes of MATHEMATICAL REVIEWS beginnning in 1990.

Page 1 — Preface, introduction, or any other matter not belonging in body of text.

Footnotes: *Received by the editor date.
Support information — grants, credits, etc.

First Page Following Introduction - Chapter Title (dropped 1 inch from top line, and centered). Beginning of Text.

Last Page (at bottom) - Author's affiliation.